CLIMATE CHANGE AND THE PEOPLE'S HEALTH

Small Books, Big Ideas in Population Health
Series Editor
Nancy Krieger

1. J. Beckfield. *Political Sociology and the People's Health*
2. S. Friel. *Climate Change and the People's Health*

CLIMATE CHANGE AND THE PEOPLE'S HEALTH

Sharon Friel

PROFESSOR OF HEALTH EQUITY AND DIRECTOR OF THE SCHOOL OF REGULATION AND GLOBAL GOVERNANCE, AUSTRALIAN NATIONAL UNIVERSITY

OXFORD UNIVERSITY PRESS

OXFORD
UNIVERSITY PRESS

Oxford University Press is a department of the University of Oxford. It furthers the University's objective of excellence in research, scholarship, and education by publishing worldwide. Oxford is a registered trade mark of Oxford University Press in the UK and certain other countries.

Published in the United States of America by Oxford University Press
198 Madison Avenue, New York, NY 10016, United States of America.

Library of Congress Cataloging-in-Publication Data
Data Names: Friel, Sharon, author.
Title: Climate change and the people's health / Sharon Friel.
Other titles: Small books with big ideas ; 2.
Description: Oxford ; New York : Oxford University Press, [2019] |
Series: Small books, big ideas in population health ; 2 |
Includes bibliographical references and index.
Identifiers: LCCN 2018025956 | ISBN 9780190492731 (hardcover : alk. paper)
Subjects: | MESH: Climate Change | Population Health |
Social Determinants of Health | Health Status Disparities
Classification: LCC RA427.8 | NLM WA 30.2 | DDC 362.1—dc23
LC record available at https://lccn.loc.gov/2018025956

9 8 7 6 5 4 3 2 1

Printed by Sheridan Books, Inc., United States of America

In dedication to the late Professor Anthony (Tony) McMichael—friend, mentor, and inspirational leader in the field of climate change and health

CONTENTS

FOREWORD

Why this book on climate change and the people's health? In keeping with the spirit of this new series of small books with big ideas (Krieger, 2018), this slim volume, authored by Sharon Friel and the second one in the roster, tackles one of the major crises of our times: climate change (Klein, 2014; Levy and Patz, 2015; Luber and Lemey, 2015; McMichael, 2015; Watts et al., 2018). With a critical eye on those who harmfully gain from what she terms a "consumptagenic society," Friel provides a cogent analysis of who and what is accountable for climate change and its predictable catastrophic and inequitable consequences, including, among others, harming people's health and exacerbating persistent, preventable, and pernicious health inequities, within and across countries, worldwide (Birn, Pillay, and Holtz, 2017; Klein, 2014; Levy and Patz, 2015).

Friel's book is a call for rapidly strengthening alliances between two sectors of scientists and advocates who have only recently begun to engage with each other: those focused on environmental and ecological health, and those focused on human health inequities (Birn, Pillay, and Holtz, 2017; Klein, 2014; Levy and Patz, 2015; Luber and Lemey, 2015; McMichael, 2015; Watts et al., 2018). Knowledge of the relevant evidence, and also both one's allies and adversaries, is crucial for making change (Krieger,

2018). So, too, is knowledge of the institutional "rules of the game" that set the terms on which societies shape people's health, for good and for bad—issues that are at the core of the first book in this series, *Political Sociology and the People's Health*, by Jason Beckfield (Beckfield, 2018).

It is, of course, daunting to take on struggles for health equity, sustainability, and social justice. It is also necessary for the people's health. Lending perspective, it is useful to recall that during the heyday of the campaign for nuclear disarmament in the 1970s and 1980s (Campaign for Nuclear Disarmament [CND], n.d.; Physicians for Social Responsibility [PSR], n.d.), Raymond Williams (1921–1988), a radical English cultural theorist and activist, wisely observed that for this struggle to be successful, it would require

(1) being clear on not just what people oppose (nuclear weapons; the threat of annihilation by nuclear war), but what people stand for (a nuclear-free, safer, peaceful, fairer world); and
(2) "making hope practical, rather than despair convincing" (Williams, 1980).

Social change, indeed, requires visionary and deep analysis combined with concrete collective organizing.

Although the threat of nuclear warfare is far from gone (CND, n.d.; PSR, n.d.), without the forward-looking practical, hopeful activism which Williams advocated, the peril would be infinitely worse. The mix of scientific evidence and activism, delineating utterly alarming risks and alternatives, commenced in the 1950s and helped bring about the 1970 landmark binding international

Treaty on the Non-Proliferation of Nuclear Weapons, currently signed by 191 countries (CND, n.d.; PSR, n.d.; United Nations Office for Disarmament Affairs, 1995). These evidence-based campaigns led to the Nobel Peace Prize being awarded in 1985 jointly to Physicians for Human Rights (founded in 1961 to educate the public about the dangers of nuclear war) and the International Physicians for the Prevention of Nuclear War, on account of their critical role in "building awareness and pressure to end the nuclear arms race" (PSR, n.d.). They also contributed to the subsequent signing, by the United States and USSR, of the 1987 Intermediate Range Nuclear Force Treaty (CND, n.d.; PSR, n.d.; US Department of State, 1987). The history of international treaties and national policies to address climate change has many parallels (Birn, Pillay, and Holtz, 2017; Klein, 2014; Levy and Patz, 2015; Luber and Lemey, 2015; McMichael, 2015; Watts et al., 2018). Awareness that "No is Not Enough" (Klein, 2017) is central to the analysis and practice of the growing social movements joining forces to build a more sustainable and equitable world, as incisively chronicled by Naomi Klein in her critical work on climate change and capitalism (Klein, 2014, 2017).

The urgent need to make hope practical keenly applies to the contemporary reality—and looming catastrophe—of global climate change, which threatens the integrity of ecosystems across our planet Earth (Birn, Pillay, and Holtz, 2017; Klein, 2014; Krieger, 2018; Levy and Patz, 2015; Luber and Lemey, 2015; McMichael, 2015; Watts et al., 2018). This is the first time, in our planet's recorded history—whether chronicled by humans in words or art, or by the fossil remains of biological beings—that actions by members of one species (not all are equally responsible) have altered the geophysical and biological terms of life itself on our globe (McMichael, 2015).

The hypothesis that rampant burning of fossil fuel could lead to rises in carbon dioxide (CO_2) that would affect planetary temperature and the global climate dates back to the early twentieth century (Levy and Patz, 2015; Luber and Lemey, 2015; McMichael, 2015; Watts et al., 2018). Also contributing is the ravaging of forests and wetlands (Klein, 2014; Levy and Patz, 2015; Luber and Lemey, 2015; McMichael, 2015; Watts et al., 2018). As I observed in an essay I wrote (Krieger, 2015) after participating in the People's Climate Change march in New York City on September 21, 2014:

> Back in 1933, Edgar Sydenstricker, in his landmark monograph Health and Environment, prepared for the US President's Research Committee on Social Trends, cogently analysed how, what he termed the geographic, social and occupational environments, both shaped population health and were reciprocally shaped by people's actions—although noting, in a passage now rueful to read, "There are some geographic conditions as yet unaltered by [people], such as the amount of rainfall, the degree of heat or cold, the topography of the land." (Sydenstricker, 1933) (p. 47)

Sydenstricker's speculative "as yet unaltered" has since shifted from imaginable to real. The terrifying fact of climate change and its health implications is now an urgent focus of scientific analysis and societal action (Levy and Patz, 2015; Luber and Lemey, 2015; McMichael, 2015; Watts et al., 2018), and is sharpening the political fights over accountability and the fierce "alternative fact" fusillade from climate change denialists and their fossil fuel sponsors (Birn, Pillay, and Holtz, 2017; Klein, 2014, 2017).

Now, nearly two decades into the twenty-first century, the science of climate change is a robust cross-disciplinary field with many foci, including "planetary health" (Watts et al., 2018). Rigorous evidence, tussled over, debated, and vetted by myriad scientists worldwide, makes clear the extraordinary disturbances and dangers that lie ahead if the CO_2 level in the atmosphere (already wreaking havoc and harm) continues to rise: extreme weather, rising oceans, imperiled ecosystems, species extinction, and, specifically relevant to humans, economic and political collapse of current societies, accompanied by vicious wars as people fight over access to and distributions of energy, water, and the material resources required to survive (Birn, Pillay, and Holtz, 2017; Klein, 2014; Levy and Patz, 2015; Luber and Lemey, 2015; McMichael, 2015; Watts et al., 2018). Important debates concern not only what types of technological innovations are needed to avert climate catastrophe, but also the skewed political and economic priorities that drive the crisis—and which must be changed for people to live in equitable and sustainable societies (Birn, Pillay, and Holtz, 2017; Klein, 2014; Levy and Patz, 2015; Luber and Lemey, 2015; McMichael, 2015; Watts et al., 2018).

All the more reason to be aware that recognition of the deep links between ecology and population distributions of health goes back centuries, if not millennia (Krieger, 2011, 2015; Rosenberg, 2012; Sydenstricker, 1933). Back in 1994, when I first articulated the ecosocial theory of disease distribution, I explicitly chose the term "eco" to signify the importance, for the well-being of people and for health equity, of real-life ecosystems and ecology, in contrast to the commonplace metaphorical use of "ecology" in social and behavioral sciences to describe networks and levels of human interactions and institutions (Krieger, 1994, 2011). Cognizant of the interplay between the biophysical conditions that enable life to exist on our planet and the different types of political and economic systems that have variously promoted health for the few versus health for all, I developed an ecosocial fractal metaphor, spanning from the macro to the nano, comprised of the inextricably bound tree (or rather bush) of life and the scaffolding of society that different groups daily seek to uphold or change (Krieger, 1994, 2011). My argument was that "this image makes clear that although the biologic may set the basis for the existence of humans and hence our social life, it is this social life that sets the path along which the biologic may flourish—or wilt" (Krieger, 1994) (p. 899).

Sharon Friel's book brings new urgency to the ecosocial metaphor. To spur the critical thinking and alliances that are needed, in the text's first chapter she beautifully provides the "101" introduction to global climate change and to global health inequities (between and within countries), and cogently explains their profound interconnections. She then explicates, in the second chapter, what she means by "consumptagenic societies," using the concrete examples of industrial food production systems and urbanization. Her third chapter orients us, the readers, to the work ahead, regarding both the evidence gaps that scientists need to address, and the political, economic, and policy changes that engaged and informed advocates (including scientists) need to propel.

Providing concrete examples of research and action that are already making a difference, Friel rises to Williams' challenge—whereby she provides the necessary insight and hope so critically needed to fuel the collective fight against climate change and for climate justice and environmental justice, and social justice and health equity more broadly. Read this book, take in its evidence and analysis, feel alarm and fury, share it with your friends, colleagues, and communities, and constructively channel the collective concern and outrage to spur research and action to protect the people's health and our planet's health, now and for future generations.

Nancy Krieger

References

Beckfield, J. (2018) *Political sociology and the people's health*. New York: Oxford University Press.

Birn, A. E., Pillay, Y. and Holtz, T. H. (2017) *Textbook of global health*, 4th ed. New York: Oxford University Press.

Campaign for Nuclear Disarmament (CND) (n.d.) "The history of CND." Available at: http://www.cnduk.org/about/history (Accessed April 19, 2018).

Klein, N. (2014) *This changes everything: capitalism vs the climate*. New York: Simon & Schuster.

Klein, N. (2017) *No is not enough: resisting Trump's shock politics and winning the world we need*. Chicago: Haymarket Press.

Krieger N. (1994) "Epidemiology and the web of causation: has anyone seen the spider?", *Social Science Medicine*, 39, pp. 887–903.

Krieger, N. (2011) *Epidemiology and the people's health: theory and context*. New York: Oxford University Press.

Krieger, N. (2015) "The real ecological fallacy: epidemiology and global climate change", *Journal of Epidemiology and Community Health*, 69, pp. 803–804.

Krieger, N. (2018) "Introduction" in Beckfield, J. (ed.) *Political sociology and the people's health*. New York: Oxford University Press, pp. xxi–xxx.

Levy, B. S. and Patz, J. A. (eds.) (2015) *Climate change and public health*. New York: Oxford University Press.

Luber, G. and Lemey, J. (eds.) (2015) *Global climate change and human health: from science to practice*. San Francisco, CA: Jossey-Bass.

McMichael, A. J. (2015) *Human frontiers, environments, and disease: past patterns, uncertain futures*. Cambridge, UK: Cambridge University Press, 2001.

Physicians for Social Responsibility (PSR) (n.d.) "History." Available at: http://www.psr.org/about/history.html (Accessed April 19, 2018).

Rosenberg, C. (2012) "Epilogue: airs, waters, places. A status report", *Bulletin of the History of Medicine*, 86, pp. 661–70.

Sydenstricker, E. (1933) *Health and environment*. New York: McGraw Hill.

United Nations Office for Disarmament Affairs (1995) "Treaty on the Non-Proliferation of Nuclear Weapons (NPT)." Available at: https://www.un.org/disarmament/wmd/nuclear/npt/ (Accessed April 19, 2018).

US Department of State (1987) "Treaty between the United States of America and the Union of Soviet Socialist Republics on the Elimination of Their Intermediate-Range and Shorter-Range Missiles (INF Treaty)." Available at: https://www.state.gov/t/avc/trty/102360.htm (Accessed April 19, 2018).

Watts, N., Amman, M., Ayeb-Karlsson, S., Belesova, K., Bouley, T., Boykoff, M., et al. (2018) "The *Lancet* Countdown on health and climate change: from 25 years of inaction to a global transformation of public health", *Lancet*, 391(10120), pp. 581–630.

Williams, R. (1980) "The politics of nuclear disarmament", in Williams, R. and Gable, R. (ed.) *Resources of hope: culture, democracy, and socialism*. London: Verso, 1989.

ACKNOWLEDGMENTS

My immense thanks goes to Dr. Emma Larking from the School of Regulation and Global Governance (RegNet), Australian National University (ANU), for her skillful editorial work, and to Janice Lee, also from RegNet, for her assistance with literature searching. I am very thankful for having had the opportunity to work with many great colleagues around the world—too many to mention here; the work presented in this book draws on some of those collaborations, in particular the World Health Organization (WHO) Commission on Social Determinants of Health; The Global Research Network on Urban Health Equity; INFORMAS; The Australian Prevention Partnership Centre; and the National Health and Medical Research Council (NHMRC) Centre for Research Excellence on the Social Determinants of Health Equity. This is a book that took far too long to write (note to self, never say yes to writing a book when starting a new position, especially as director of a school within a university!). And I am deeply thankful to Chad Zimmerman and Oxford University Press for their incredible patience with me. I also thank Professor Nancy Krieger of Harvard University for her kind invitation to do this book and also for her patience. Finally, a big thanks to David for all his support—he didn't see me over many weekends, and never complained.

INTRODUCTION

Climate change threatens humanity and the planet on which we live. Social inequities, including startling variance in the health outcomes that different population groups enjoy, also pose a threat to humanity, although less directly. Together, the scale of devastation these threats pose is unprecedented, but wholesale destruction is not inevitable. Humanity can and must act to prevent catastrophic climate change and redress egregious global health inequities. We must act now. This book outlines some of the steps necessary to move from denial and inertia toward effective mobilization.

The book makes three key contributions to the current literature and understanding of climate change and health inequity. First, it describes how climate change interacts with and exacerbates existing health inequities. While there has been considerable discussion to date of the impact that climate change will have on human health (Costello et al., 2009; McMichael, 2009; Whitmee et al., 2015), little sustained attention has been directed at its significance for health *inequity*. Second, the book introduces the concept of a "consumptagenic system." This is an integrated network of policies, processes, governance, and modes of understanding that fuel unhealthy, environmentally destructive production and consumption. Third, the book argues that mobilizing effective action on climate change and health inequity requires a systems approach. Systems science developed as a way of organizing and analyzing complex information in a manner attentive to the overall operation of systems and to the dynamic interaction of variables within them. It is an approach that can be adapted in response to complex social problems such as climate change and health inequity.

The book calls for interdisciplinary responses that employ knowledge and analytical tools from across the sciences, social sciences, and even humanities. It advocates as well for intersectoral engagement across policy domains. The book's intended audience is broad: it will assist epidemiologists, public health scholars, and policymakers to understand the effect of climate change on health inequities. Scientists and policymakers whose work focuses on climate change will also benefit from the insights that the book provides to the human face and health impacts of climate change. The book is aimed as well at ecological and progressive economists who understand that the current economic system of globally integrated markets, and reliance on market mechanisms for the provision and protection of public goods, cannot continue. This economic system drives the consumptagenic system, producing a double burden of climate change and health inequity. The evidence of this, provided in the book, adds weight to the progressive economic case for changing the market-based system. More generally, I hope that scholars and policymakers from diverse backgrounds will be persuaded by the arguments presented here of the pressing need to respond to both climate change and health inequity, and to do so using the tools of systems science and with a clear understanding of how the consumptagenic system operates.

A Critical Juncture in Human History

> . . . We are confronted with the fierce urgency of now. In this unfolding conundrum of life and history, there is such a thing as being too late. . . . We may cry out desperately for time to pause in her passage, but time is adamant to every plea and rushes on. Over the bleached bones and jumbled residues of numerous civilizations are written the words, "Too late."
>
> —DR. MARTIN LUTHER KING, Jr.,
> *"Beyond Vietnam—A Time to Break Silence"*

Human activity is the primary cause of the changes in the earth's climate conditions since the eighteenth century that jeopardize its capacity to sustain life (Steffen, Crutzen, and McNeill, 2007). Humanity has produced and consumed vast quantities of fossil fuels and destroyed numerous animal habitats and species. We have used up almost half of the world's land surface to feed ourselves, while simultaneously concentrating large populations of people into concrete jungles. The attendant disruption to the Earth system is profound. An accumulation of carbon dioxide in the atmosphere is warming the planet. The acceleration of melting ice and river run-off is increasing the amount of freshwater that enters the oceans, making them less saline. One result is that thermohaline circulation, which moves warm, saline waters from the tropics to the north before looping back, could collapse. If this happens, the North Atlantic would cool quickly, creating dramatic but varied changes to temperature and precipitation across the Northern and Southern Hemispheres.

While the disruption caused by climate change to life-supporting environmental systems will affect everyone, it will have the greatest—and generally earliest—impact on the poorest and most disadvantaged populations and nations. This will compound existing social inequities within and between nation states. As the epidemiological evidence demonstrates, these social and global inequities are causes of, and further entrenched by, health inequities.

Social inequities, and the health inequities that flow from and deepen them, are not inevitable or immutable. Political, economic, and cultural norms operate to justify how power, prestige, wealth, and other resources are socially distributed. Such norms also play a role in how resources are generated and valued. Societies in which unequal distributions of resources are normalized create structural inequities in the daily conditions of people's lives (Navarro, 2000). We now live in a global society characterized by extreme wealth and income inequity within and between nations (Inequality.Org, 2018). The country in which you are born and the wealth of the family you are born into, as well as the color of your skin, your gender, and your ethnic and cultural origins, influence your daily living conditions. These include your access to quality education and healthcare, to sufficient nutritious food, the conditions of your work and leisure, the quality of your housing and built environment, and your social relations. Together, these factors affect human health and produce health inequities (Commission on Social Determinants of Health, 2008; Marmot et al., 2008). In the field of social epidemiology, the political, economic, and cultural norms and practices that structure social life and influence health outcomes are referred to as the *social determinants of health* (Commission on Social Determinants of Health, 2008) (Box I.1).

There has been increasing research and policy attention to the social determinants of health and health inequities (Commission on Social Determinants of Health, 2008; World Health Organization [WHO], 2015a, 2016b). As noted earlier, there is also a growing literature on the impacts of climate change on human health. However, little is documented specifically in relation to the connections between climate change and the social determinants of health and health inequities. Making

Box I.1

Distinguishing Health Inequality and Health Inequity

Health *inequality* can be defined as the divergence in health outcomes experienced by different social groups and nations. Health *inequity* refers to those differences in health outcomes that are a consequence of unjust social arrangements and, as such, are avoidable and may be remedied. The starting point of this book is that if there is no biological reason for differences in health, then they are not inevitable. If differences in health are not inevitable, then a large part of the failure to avoid or remedy them is likely to be found in political and social arrangements and constitutes a failure of social justice.

these connections is critical in order to effectively harness the political momentum that is currently building in response to health inequities on the one hand, and climate change on the other.

Where There's a Will, There May Be a Way: An Overview of Global Action on the Social Determinants of Health and on Climate Change

The Commission on Social Determinants of Health (the Commission) was established in 2005 by the then WHO Director-General J.W. Lee to gather evidence on what could be done to achieve better and fairer health outcomes globally (WHO, 2018). The Commission was a global network of policymakers and researchers (including the author of this book), as well as civil society organizations, committed to tackling health inequities through a focus on their social, political, and economic causes. In 2008, the Commission released its final report, "Closing the Gap in a Generation: Health Equity through Action on the Social Determinants of Health." The report's recommendations target public policies and social arrangements that have contributed to "gross" health inequities both within nation states and globally (Commission on Social Determinants of Health, 2008). They emphasize the need to tackle social structures that facilitate or produce inequitable distributions of power, money, and resources—all of which lead to inequities in health outcomes (Commission on Social Determinants of Health, 2008).

The report argues that the knowledge and evidence base exists to achieve health equity within a generation. It describes this achievement as an ethical imperative and a matter of social justice, but the authors acknowledge the many hurdles in its way. They call for an approach going well beyond the traditional public health focus on the health sector, in particular health services. What is needed instead, the report urges, is the involvement not just of health ministries but of "the whole of government." In addition to government actors, the report notes the need for the inclusion of and actions by "civil society and local communities, business, global fora, and international agencies" (Commission on Social Determinants of Health, 2008).

The 2008 Commission report also notes the "profound" impacts of climate change "for the life and health of individuals and the planet." It recognizes the pressing need to "bring the two agendas of climate change and health equity together" (Commission on Social Determinants of Health, 2008). Since its publication, however, the connections between climate change and health inequities have received scare mention in the momentum steadily building to advance recognition and action on the social determinants of health within the field of epidemiology and, importantly, at the policy level.

Much of the related global policy action has centered on the *Rio Political Declaration on Social Determinants of Health*, adopted in 2011 by 124 WHO member states and endorsed the following year by all 194 members (WHO, 2011, 2016b). The first paragraph of the Declaration expresses "the determination" of member states "to achieve social and health equity through action on the social determinants of health and well-being by a comprehensive intersectoral approach." Yet the Declaration addresses environmental concerns in a way that effectively puts them aside. It notes that the social conditions affecting health inequities include the environment in which people live (para 6). In the Declaration, member states pledge to address a range of challenges, which include "ensuring food and nutritional security . . . [and] protecting environments" (para 8). They also express their commitment to developing health policies that address the environmental determinants of health and health inequities (para 13.2.i). No mention is made in the Declaration of climate change. Nor is there mention of the wholesale threats that environmental degradation poses to life on earth, or of the underlying causes of these threats (WHO, 2011, 2016b). Subsequent efforts to monitor and implement the Rio Political Declaration have also been silent on the topics of climate change and other major environmental issues, as well as the role they play in exacerbating health inequities (WHO, 2016b).

In contrast to this silence in public health, there has been greater recognition of the connections between inequity and climate change in the development field. The United Nations Development Programme's (UNDP) Human Development Report 2011 is titled *Sustainability and Equity: A Better Future for All.* Noting the advances in development that had been made over previous decades, the report cautions: "continuing failure to reduce . . . grave environmental risks and deepening inequalities threatens . . . to reverse the global convergence in human development" (UNDP, 2011). It explicitly highlights the connections between inequity and vulnerability to climate change and other environmental harms. The report synthesized knowledge that at that time was generating enthusiasm for and work on the UN's Agenda for Sustainable Development.

The UN launched the "2030 Agenda for Sustainable Development" in 2015 and set out 17 Sustainable Development Goals (SDGs). These are the successors to the UN's earlier Millennium Development Goals. The Agenda has been characterized by some researchers as providing an enabling framework for advancing the social determinants of health approach (Donkin et al., 2018). As the title makes clear, the SDGs overtly flag the shift in development thinking and the aspirations signaled by the 2011 Human Development Report. This reflects a change from development as a goal that is desirable regardless of its associated costs toward the pursuit of development that is desirable only if it is also environmentally sustainable. The 17 SDGs are further subdivided into 169 targets that constitute an action agenda for the years 2015–2030. Taken together, the UN describes the goals as being designed to promote peace and prosperity, eradicate poverty, and "heal the planet" (United Nations, 2015).

The Agenda for Sustainable Development demonstrates that the significance of structural influences on health and well-being is increasingly acknowledged. The third Sustainable Development Goal (SDG 3) focuses explicitly on good health and well-being, with the aim to "ensure healthy lives and promote well-being for all at all ages" (UN, 2015). While no specific mention is made of the importance of social determinants for health outcomes, either within SDG 3 or elsewhere in the Sustainable Development Agenda, almost all of the SDGs have implications for health, and on their face at least, they capture many important determinants of health. Thus there are goals focusing on poverty (SDG 1), hunger (SDG 2), education (SDG 4), gender equality (SDG 5), clean water and sanitation (SDG 6), affordable and clean energy (SDG 7), inequality (SDG 10), sustainable cities and communities (SDG 11), responsible consumption and production (SDG 12), climate action (SDG 13), life below water (SDG 14), life on land (SDG 15), and peace, justice, and strong institutions (SDG 16).

According to the UN, the Sustainable Development Agenda differs from its predecessor, the Millennium Development Agenda, by recognizing "that tackling climate change is essential for sustainable development and poverty eradication" (UN, 2018). The Sustainable Development Agenda is certainly attentive to environmental concerns in general and to climate change in particular. Most of the goals are relevant to climate change, and environment-related factors directly relevant to it are the explicit focus of Goals 12–15: "ensure sustainable consumption and production patterns" (SDG 12); "take urgent action to combat climate change and its impacts" (SDG 13); "conserve and sustainably use the oceans, seas and marine resources for sustainable development" (SDG 14); and "protect, restore and promote sustainable use of terrestrial ecosystems, sustainably manage forests, combat desertification, and halt and reverse land degradation and halt biodiversity loss" (SDG 15).

The Sustainable Development Agenda marks a key moment in global policymaking by recognizing that responses to climate change (and other environmental issues) cannot be siloed away from other major concerns for human progress, including advances in health outcomes. However, I will argue in

Chapter 3 that the SDGs are also somewhat of a diversion insofar as there is a need to tackle the underlying common drivers of climate change and health inequities. For the purposes of this Introduction, it is necessary to note only that while the 2011 UN Development Report explicitly tied equity and environmental sustainability together, characterizing them as interconnected and with the potential to be mutually reinforcing (UNDP, 2011), the SDGs are much less centrally concerned with inequity or with redressing inequality. They specifically include the goal of reducing inequality within and among countries (SDG 10), but even so, they do not genuinely bring together inequity and climate change concerns. This is because they do not address the amplifying impact that climate change will have on inequity. Moreover, the actual targets that are designed to respond to inequality are very weak. This is discussed in greater detail in Chapter 3.

In the same year that the Agenda for Sustainable Development was adopted, negotiators met in Paris for the 2015 UN Climate Change Conference (COP21). The outcome of COP21—the Paris Agreement—has been ratified by 126 countries and entered into force in November 2016. It aims to strengthen the global response to climate warming by keeping the overall temperature rise during this century to "well below" 2°C in comparison with pre-industrial levels. The Agreement further aspires to pursue efforts to keep the rise below 1.5°C. However, under the Agreement, action on reducing greenhouse gas emissions is not scheduled to begin until 2020. This includes action to mitigate the impacts of climate change, to adapt to its effects, and to provide enabling finance (International Food Policy Research Institute, 2017, United Nations Framework Convention on Climate Change [UNFCCC], 2018). I discuss the technicalities of the connection between climate change and climate warming, and the role of greenhouse gas emissions, in Chapter 1.

As is apparent from COP21, achieving a global political consensus on climate change is proving exceptionally difficult. This is despite the fact that political will exists in some quarters to respond immediately and proactively to the threats it poses. Barriers to consensus are exacerbated by inequities in negotiating power between states, and also between states and non-state actors. These barriers are further exacerbated by differences in who benefits from the industrial processes that drive climate change and those most vulnerable to its impacts. I discuss this further in Chapters 1 and 2.

In the final part of Chapter 3 I return to the question of what political will exists to tackle climate change and health inequities. Alarmingly, while writing this book, the Trump administration pulled out of the Paris Agreement. However, it is encouraging that there has been a global outcry and that leaders in countries across the world have re-pledged their commitment to the Agreement. Within the United States, more than 1,200 states, cities, and business leaders have banded together to continue working toward meeting the Agreement's targets—networks of hope in the context of institutional resistance. Here we see the redemptive possibilities of complexity—a topic I take up in Chapter 3 when discussing how existing political will might be harnessed and expanded in response to climate change and health inequity.

The Story

Chapter 1 commences with a brief status update on global environmental change. It introduces Crutzen's concept of the "Anthropocene"—a new geological epoch in which the scale and impact of human activity is destabilizing the planet's capacity to support viable human communities—and discusses the "planetary boundaries" framework for assessing environmental threats. Against this more general backdrop, the book's focus is on climate change.

In Chapter 1 I explain what climate change is, the role of human activities in driving climate change, and some of its most significant impacts, including extreme weather events and altered hydrological systems, sea level rise, increased oceanic acidity, species extinction, and food insecurity. The chapter then turns to a discussion of some of the key justice issues raised by climate change, including issues relating to causal responsibility, future development rights, the distribution of climate change harms, inequity in the negotiating positions of states, and intergenerational inequity.

Next, a status update on current health inequities is provided. While average life expectancy has been increasing globally since 1950 (with the exception of the 1990s), there are marked inequities in life expectancy between states, and within states between different population groups. The effects of communicable diseases—which continue to ravage some parts of the world despite having largely been eradicated in others—play a role in this, as does undernutrition. At the same time, burgeoning rates of noncommunicable diseases, including obesity and respiratory diseases, are producing health inequities. A decade or two ago, these inequities would have been analyzed primarily in terms of individual behavior and risk factors. Now, however, the role of political, economic, and cultural norms and of the social institutions and structures that they establish and justify is increasingly recognized.

Chapter 1 canvasses the shift in epidemiological focus from individual behavior to social determinants of health, discussing the significance for health of material resources, psychosocial resources, and social inclusion. This part also discusses the evolution of environmental epidemiology—concerned with environmental health risks and inequities—and from this starting point, the shift to eco-social approaches and eco-epidemiology, which are ecologically informed, and analyze how interactions between physical and social environments influence health outcomes. While eco-epidemiologists have begun to research the influence of climate change on health, this research has not yet considered in depth the influence of social systems.

The final section of Chapter 1 provides an overview of how climate change will exacerbate existing health inequities, focusing on the health implications of the significant climate change impacts outlined earlier: extreme weather events, rising sea levels, heat stress, vector-borne diseases, and food insecurity.

While Chapter 1 provides an account of the social determinants of health inequity and the role of climate change in amplifying health inequities, Chapter 2 identifies a loop or system in which some of the key drivers of health inequity also fuel climate change, which in turn fuels further inequity. This process is based on excessive production and consumption—it constitutes a consumptagenic system.

Chapter 2 tracks the evolution of the consumptagenic system through the globalization of a market-based and fossil fuel–dependent economic system. It describes the addiction of this system to growth and to forms of consumption that are highly polluting. The second and third parts of the chapter focus on and describe the roles of an industrial food system and urbanization as two central cogs in the consumptagenic system that is pushing our planet toward irreparable destabilization.

The second part of Chapter 2 explains the character of the industrial food system and charts the history of its growth and eventual dominance over other food systems. The impact that the industrial food system has on the relative accessibility and affordability of unhealthy food, as opposed to healthy food, is considered, as well as its role in burgeoning rates of noncommunicable diseases. This section discusses the ways in which the industrial food system contributes to the social conditions of people's lives to create and exacerbate diet-related health inequities. The section concludes with a summary of the environmental, climate-change, and health-related impacts of the industrial food system.

The third part of Chapter 2 considers the implications of urbanization—from a world in which just 1 in 10 people lived in cities at the dawn of the twentieth century to the current situation in which more than half of the global population are urban dwellers. The chapter posits that while urbanization has brought greater prosperity and opportunities for many, it is fueling modes of production and consumption that greatly exacerbate climate change. At the same time, it has created and compounded health inequities. Poor urban living conditions are implicated in the spread of communicable diseases, particularly in informal settlements in the developing world, but also throughout many low- and middle-income countries. Urban living conditions are also playing a central role in the rapid global growth of noncommunicable diseases. Climate change will magnify existing health inequities associated with urban living. In this section I consider how climate change will interact with urban health inequities via rising temperatures, sea level rise, and inland flood risks.

Chapter 3 looks forward. It begins by calling for a wider conceptual and methodological toolkit among public health researchers—one that includes systems science as well as a focus on policymaking processes. I explain what is distinctive about a systems approach and why resorting to it is imperative, both in research and in practice. I also call for public health researchers to pay more attention to political and policy processes. Rather than engaging solely with a problem that has been identified and thus being drawn into reductionist pathologizing, public health researchers need to seek and work with evidence of how to mobilize change.

The next part of the chapter proposes a policy vision. It considers how progressive policy systems might be created and deployed to reign in consumption and sets forth some options for intersectoral action designed to achieve greater equality, environmental sustainability, and health equity. The focus here is on reducing material inequity to produce systemic effects that are preconditions for responding to climate change, as well as for the achievement of health equity. In this part I also review some of the existing policy responses to climate change and health inequity discussed earlier in this introduction. While admitting that the proposed policy vision and actions to achieve it will confront stubborn resistance,

I conclude that change is possible—and that what seems like cause for despair might also represent cause for hope.

Issues the Book Does Not Address

There are major threats to healthy human and planetary flourishing that are ignored or discussed only superficially in this book. Although critically important, climate change is just one environmental process fueled by human activity that is destabilizing planetary life-support systems; others include stratospheric ozone depletion and changes in biochemical flows. The primary focus of this book, however, is on climate change.

Although the book engages with some key social determinants of climate change and health inequity by identifying the role played by the consumptagenic system, there are other contributors to climate change and health inequity that the book does not address. Most notable among these is population growth: the questions of how many people the earth can sustainably support and how population growth can or should be limited in line with this are not addressed here.

While the role played by fossil fuels in powering the consumptagenic system is discussed, the book does not address the environmental and social issues associated with transitioning to alternative energy sources. The relative costs and benefits of power sources such as nuclear, solar, wind, and waves are contentious. There are, moreover, vested interests and strong lobby groups seeking investment in nuclear power and advocating for less reliance on other sources. This book's account of the consumptagenic system and the argument that we need systems thinking in order to formulate effective policy responses to it provide tools for thinking through these issues, although the book does not address them directly.

1 CLIMATE CHANGE, GLOBAL JUSTICE, AND HEALTH INEQUITIES

Status Update on Global Environmental Change

It's the end of the world as we know it. The "Anthropocene" concept proposed by Crutzen (2002) suggests that, in the nineteenth century, we entered a new geological epoch, signaling a profound shift in the relationship between humans and the natural environment. In the postglacial Holocene period, which lasted more than 10,000 years, the Earth system accommodated humans and enabled them to survive and flourish (Box 1.1). In the Anthropocene, the geological scale and impact of human activity is rivaling natural forces and disrupting the Earth system's ability to provide an accommodating environment for further human development (Crutzen, 2002; Steffen, Crutzen, and McNeill, 2007).

Building on this idea of now living in the Anthropocene, the planetary boundary framework, developed by Steffen and colleagues (2015), provides an analytical tool for assessing the impact of human activity on the Earth system and identifying risks of destabilization. The framework incorporates nine processes

Box 1.1

Defining the Earth System

Steffen, Crutzen, and McNeill define the "Earth system" as "the suite of interacting physical, chemical and biological global-scale cycles and energy fluxes that provide the life-support system for life at the surface of the planet." They point out that "feedbacks *within* the Earth system are as important as external drivers of change, such as the flux of energy from the sun." They also emphasize that humans are not external to the Earth system "but rather an integral and interacting part of [it]" (Steffen et al., 2007).

influenced by humans that are critical to the functioning of the Earth system (Box 1.2).

The framework identifies threshold levels below which human activity is unlikely to destabilize the Earth system. These "planetary boundaries" "delineate 'a safe operating space' for global societal development" (Steffen et al., 2015). Of concern is the fact that the impacts of human activity have exceeded planetary boundaries in respect of four processes: climate change, biosphere integrity, biogeochemical flows, and land system change (Figure 1.1).

A Focus on Climate Change

In this book I am concerned primarily with climate change. Along with biosphere integrity, climate change represents a "core"

Box 1.2

Nine Processes Caused or Influenced by Human Activity That Are Critical to Functioning of the Earth System

The nine processes identified by Steffen and colleagues are climate change; (the potential production of) novel entities; stratospheric ozone depletion; atmospheric aerosol loading; ocean acidification; biochemical flows; freshwater use; land-system change; and biosphere integrity. "Novel entities" are new or modified substances or life forms that may produce destabilizing geophysical and/or biological effects (Steffen et al., 2015).

planetary boundary because of its centrality to the functioning of the Earth system and its interrelationship with other critical processes (Steffen et al., 2015). Climate change involves changes in weather patterns caused by an increase in the earth's average temperature, which is itself caused by a proliferation within the earth's atmosphere of gases such as carbon dioxide (CO_2), methane, and nitrous oxide. These are known as "greenhouse gases" (GHGs) because they absorb heat from the sun and prevent it from leaving the earth's atmosphere, creating a greenhouse effect that makes the earth warmer. While GHGs are a natural part of the atmosphere, their concentration has increased rapidly as a result of human activities, initially triggered by the Industrial Revolution of the late

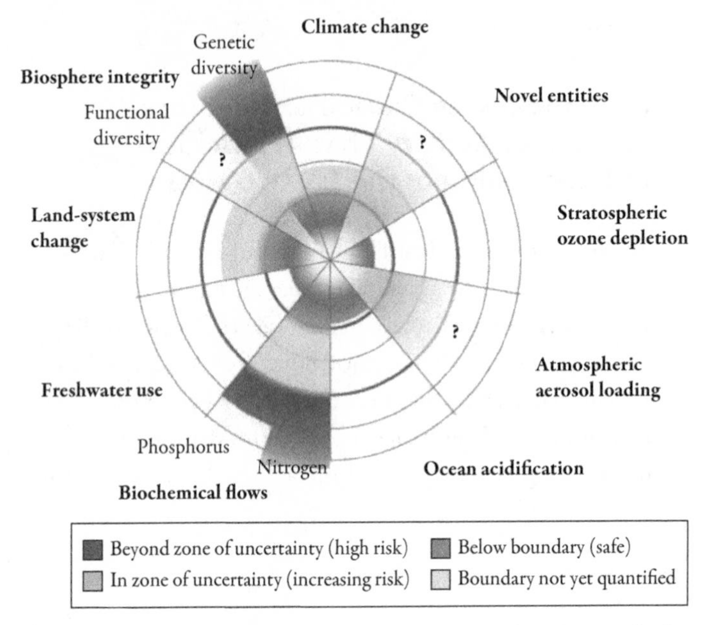

Figure 1.1 Status of nine planetary boundaries, circa 2015. Source: Steffen et al. (2015), used under the terms of the Creative Commons Attribution 4.0 license.

eighteenth and early nineteenth centuries, and with the most pronounced acceleration dating from around 1950 (Steffen et al., 2007, 2015).

An Intergovernmental Panel on Climate Change (IPCC) was established in 1988 by the World Meteorological Organization (WMO) and United Nations Environment Programme to provide evidence-based assessments of the existence and likely impacts of climate change. The IPCC's most recent analysis warns: "Warming of the climate system is unequivocal, and since the 1950s, many of the observed changes are unprecedented over decades to millennia" (IPCC, 2014b). The IPCC calculates that between 1880 and 2012, the earth warmed by just under 1°C (0.85°C), based on globally averaged land and ocean surface temperatures (IPCC, 2014b).

This may not sound like a significant increase, but even small increases in average temperatures have destabilizing effects on the Earth system. Another cause for concern is that the rate of warming is accelerating; the IPCC notes that "[e]ach of the last three decades has been successively warmer at the Earth's surface than any preceding decade since 1850" (IPCC, 2014b). Complicating matters further, although evidence of global warming is robust when calculated over decades, there is significant variability in average temperatures calculated over shorter time frames (IPCC, 2014b).

Anthropogenic Greenhouse Gas Emissions

During the Industrial Revolution, humans made great technological advances that facilitated large-scale manufacturing, radical innovations in transport, and much higher yields from agriculture. These advances came, however, at a steep price: they depended on a huge expansion in the use of fossil fuels such as coal, oil, and gas, and this use has added significantly to the levels of CO_2 and other GHGs in the atmosphere. There is consensus among international climate scientists that warming of the earth's atmosphere since the Industrial Revolution is mostly due to human activities (IPCC, 2014b). The IPCC notes that atmospheric concentrations of key GHGs—CO_2, methane, and nitrous oxide—have increased to levels "that are unprecedented in at least the last 800,000 years" (IPCC, 2014b). Consistent with the acceleration in global warming highlighted by the IPCC, much of this increase has occurred since 1950, as mentioned earlier, and almost half since the late 1970s (IPCC, 2014b; Steffen et al., 2007).

CO_2, the dominant GHG, persists in the atmosphere for many decades, some of it for centuries. CO_2 is also absorbed by the ocean, causing acidification. Its atmospheric concentration of 403 parts per million (ppm) in 2017 is 45% higher than the pre-industrial level (WMO, 2017). This increase is primarily attributable to emissions from burning and using fossil fuels and secondarily to emissions related to changes in land use, including deforestation (see Figure 1.2).

The additional warming effect of nitrous oxide and methane—both of which have much greater warming effects per unit—have raised the CO_2-equivalent concentration in the atmosphere well above 400 ppm. This means that humanity has now emitted almost 90% of the total CO_2 equivalent that the atmosphere is capable of absorbing while ensuring it remains "likely" that global temperature increases from pre-industrial levels will be capped below 2°C this century (IPCC, 2014b).[1] Preventing temperature rises beyond this will require what the IPCC calls "a stringent

1. The IPCC uses standard terminology to express the likelihood of certain events: "The likelihood, or probability, of some well-defined outcome having occurred or occurring in the future [is] described quantitatively through the following terms: Virtually certain >99% probability of occurrence; Very likely >90% probability; Likely >66% probability; More likely than not >50% probability; About as likely as not 33 to 66% probability; Unlikely <33% probability; Very unlikely <10% probability; Exceptionally unlikely <1% probability." (IPCC, 2014b)

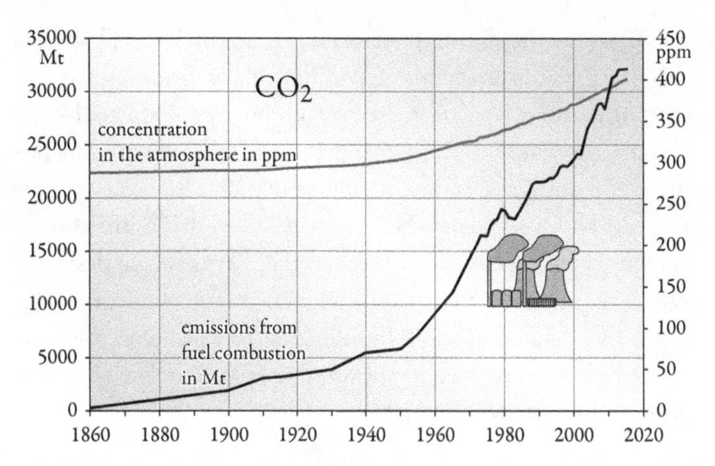

Figure 1.2 Development of global carbon dioxide (CO_2) emissions from fuel combustion and CO_2 concentration in the atmosphere. Mt, megatonnes. From Quaschning (2016), used under the terms of the Creative Commons Attribution 4.0 license.

mitigation scenario" (IPCC, 2014b), or one in which *global annual GHG emissions are reduced to zero* well before 2100 (IPCC, 2014b).

The measure 2°C is widely regarded as a critical climate change threshold, but its status as such is arbitrary from a scientific point of view (Titley, 2017). If the aim is to prevent destabilization of the Earth system and ensure "a safe operating space" for global social development, as Steffen and colleagues articulate it, or in Titley's account, to keep climate change within the bounds of "what the world ha[s] experienced in the relatively recent geological past to avoid catastrophically disrupting both human civilization and natural ecosystems" (Steffen et al., 2015; Titley, 2017), then there is no guarantee that limiting temperature rises to 2°C will achieve this. In 1990, the Stockholm Environmental Institute argued that capping global temperature rises at 1°C above pre-industrial levels was the safest option. Recognizing the difficulty of achieving political consensus and action in support of such a limit, however, it suggested 2°C as "the next best" target (Titley, 2017). The current consensus among climate scientists is that Earth system destabilization risks increase significantly with any rise in temperature beyond 1°C. As noted in the Introduction to this book, the Paris Agreement aims to keep temperature rises "well below" 2°C this century, and ideally below 1.5°C.

Environmental Impacts of Climate Change

The environmental impacts of climate change vary between regions, but climate change is already affecting natural processes and humans' interactions with these on all continents and across the oceans (IPCC, 2014b). It is associated with an increase in extreme weather events, including heatwaves, droughts, storms, and flooding; alterations in hydrological systems that are undermining water security; sea level rise; ocean acidification; species extinction; and food insecurity. The future impacts of climate change are difficult to predict with specificity but will reflect the influence of GHGs already emitted as a result of human activities, future anthropogenic emissions, secondary causes such as carbon emissions that may be released by melting arctic permafrost as the earth warms, and natural climate variability (IPCC, 2014b).

Extreme Weather Events and Alterations in Hydrological Systems

Since about 1950, global warming has caused an overall decrease in the number of cold days and nights and an increase in the number of warm days and nights, along with decreased cold-temperature extremes and increased hot-temperature extremes (IPCC, 2014b). The IPCC considers it "likely" that the frequency of heatwaves has increased in large parts of Europe, Asia, and Australia as a result of climate change and "that human influence has more than doubled the probability of . . . heat waves in some locations" (IPCC, 2014b). Fourteen of the fifteen hottest years on record have occurred since 2000, and in June 2016 the combined average temperature over global land and ocean surfaces was 0.9°C above the twentieth-century average (Jackson et al., 2015). The majority of the world's land surface had warmer to much-warmer-than-average temperatures during June 2016, with the largest temperature departures observed across much of north-central Russia, the Russian Far East, and northern Australia, where temperature departures were 3°C or higher. The only land area with cooler-than-average conditions during June 2016 was central and southern South America (National Oceanic and Atmospheric Administration [NOAA], 2016).

In the future, as global mean temperatures increase, the IPCC considers it "virtually certain" that there will be more hot- and fewer cold-temperature extremes in most places. While occasional cold-weather extremes will continue to occur, heatwaves are "very likely" to increase in frequency and duration (IPCC, 2014b). In the absence of aggressive efforts to limit GHG emissions, climate change scenarios canvassed by the IPCC project an average warming of 4–7°C over much of the global land mass by the end of the twenty-first century (IPCC, 2014b). If average change of this magnitude occurs, the hottest days will far exceed present maximum temperatures.

Humanity's ability to live and flourish on Earth is dependent on the availability and reliability of water supplies (Allan, 2014). As the planet warms, changes in precipitation patterns and melting snow and ice are affecting hydrological systems, altering the quantity and quality of water resources, and doing so with a high degree of regional variability (IPCC, 2014b). This was demonstrated in 2014 when severe drought conditions across California led to water use restrictions and wildfires. At the same time, parts of Europe were affected by torrential rainfall and flooding, causing widespread damage to infrastructure. According to the IPPC, heavy precipitation events have, on average, increased rather than decreased over the earth's land mass, and many regions are facing increasing risks of extreme precipitation and flooding (IPCC, 2014b). At the other end of the spectrum are increases in the frequency, duration, and severity of droughts in some areas. This has been found to be significant in Africa, Eastern Asia, the Mediterranean region, and Australia, while the Americas and Russia have experienced a decrease in the frequency, length, and severity of droughts (Spinoni et al., 2014). Figure 1.3 shows an overview of drought hot spots in the last 60 years. Overall, the IPCC predicts that climate change will reduce surface and ground water supplies in most dry subtropical regions, inflaming competition for this increasingly limited resource (IPCC, 2014b).

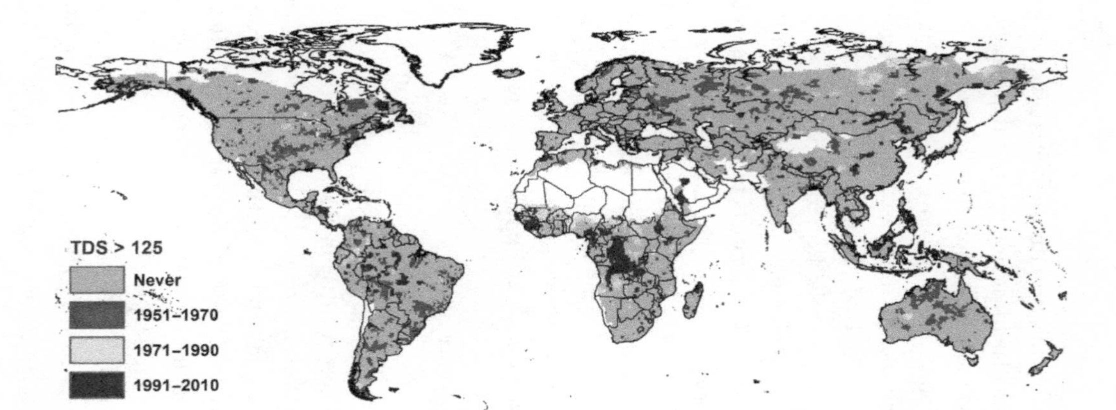

Figure 1.3 Drought hot spots in the periods 1951–1970, 1971–1990, and 1991–2010. Source Spinoni et al. (2014), used under the terms of the Creative Commons Attribution 4.0 license.

Sea Level Rise

Sea level rise is one of the most threatening consequences of global warming, in particular for low-lying coastal areas, which are expected to become more vulnerable to storm surges, flooding, loss of freshwater supplies, and land loss. As these areas often have dense populations, important infrastructure, and high-value agricultural and biodiverse land, the impacts will be far-reaching, as I discuss further later in this chapter and in Chapter 2. Between 1901 and 2010, the global mean sea level rose by about 0.19 m (IPCC, 2014b). The IPCC predicts that sea levels will continue to rise, and at an increasing rate (IPCC, 2014b). One factor in this acceleration is that ice loss from the Greenland ice sheet has been more extensive in recent years than earlier modeling predicted (Khan et al., 2015). Depending on the rate of future climate change, the global mean sea level could rise between 0.26 and 0.82 m by 2100 (Cazenave et al., 2014; IPCC, 2014b).

Increasing Ocean Acidity

Emissions of CO_2 since the Industrial Revolution have had a huge impact on oceanic acidity. Since the early nineteenth century there has been a 26% increase in acidity, and the IPCC's recent estimates suggest that if current trends persist, there could be an increase in oceanic acidity by the end of the twenty-first century of as much as 109% (IPCC, 2014b). Such an alteration in the marine environment would likely have devastating results both for ocean organisms and for the people who depend on them (Johnson and White, 2014). Shellfish and shelled zooplankton are particularly vulnerable to increased acidity, which impedes their ability to develop robust shells. As important food sources for fish, the impact of acidification on shellfish and zooplankton has significance throughout the entire marine food chain. Rising acidity is also damaging coral reef development, with flow on effects for marine biodiversity as well as for the protection of coastal environments (NOAA, 2017).

Species Extinction

Beyond the marine environment, many plant and animal species face extinction as a result of climate change and related changes in the Earth system, being unable to adapt quickly enough to the new conditions (IPCC, 2014b). The IPCC points out that there are historical precursors for this, although not caused by human activities: "natural global climate change at rates lower than current anthropogenic climate change caused significant ecosystem shifts and species extinctions during the past millions of years" (IPCC, 2014b). Even relatively minor variations in average global temperatures are likely to disrupt or destroy many natural environmental assets, species, and ecological processes (Rockstrom et al., 2009). Biologists predict that as a result of habitat destruction alone, two-thirds of all species on earth could be lost by the end of this century (Chivian, 2002).

Food Insecurity

If unchecked, climate change will place tremendous and perhaps overwhelming stress on the availability of food. Most agriculture worldwide relies on rainfall and is vulnerable to changes in precipitation, as well as to temperature increases and extremes. As noted earlier, without mitigation efforts that go well beyond those currently in place, the IPCC predicts an increase in average temperatures, by comparison with averages for 1850–1900, of 4–7°C over much of the planet's land mass by the end of this century (IPCC, 2014b). Unpredictable events such as major volcanic eruptions could push temperatures even higher. While some cold regions will benefit as new areas open up for crops and growing seasons lengthen, overall, the expected impacts are negative and include crop damage and lower yields, species extinction, animal ill health and mortality, and erosion and degradation of soils (Grist, 2015).

Some projections suggest that 70% more food will need to be produced to feed an anticipated global population of 9.1 billion people by 2050 (Food and Agriculture Organization [FAO], 2009; United Nations Department of Economic and Social Affairs [UNDESA], 2011). Yet the IPCC's most recent assessments suggest a global fall in average crop yields of greater than 5% are "more likely than not" after 2050 and "likely" by the end of the century (IPCC, 2014a). Already, IPCC assessments show that climate change is affecting food production and having an overall negative impact on agricultural yields (IPCC, 2014b) (Figure 1.4).

It is not just agriculture that is important for ensuring the availability of nutritious food. Fisheries and aquaculture represent an important source of nutrition for many of the world's peoples, and the Food and Agriculture Organization (FAO) regards the protection of fisheries and aquaculture as essential for ensuring global food security (FAO, 2012, iii). Climate change is altering the distribution of marine species and, as discussed earlier, is affecting marine biodiversity. It is predicted that climate change will reduce wild fish stocks and limit the productivity of aquaculture through changes in winds, higher water temperature, and increased ocean acidity (Mendler de Suarez et al.,

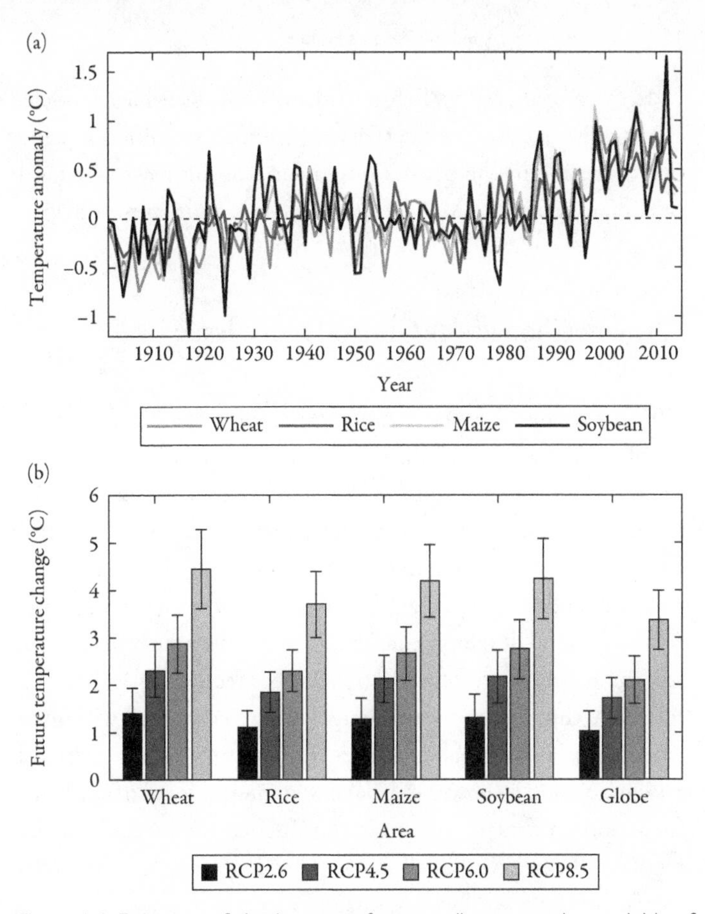

Figure 1.4 Estimates of the impacts of recent climate trends on yields of four major crops. Note: Mean annual temperature changes over time. (A) Historically observed temperature anomalies relative to 1961–1990 for global growing areas of four individual crops. (B) Future projected temperature changes (2071–2100 in comparison with 1981–2010 baseline) of four crop-growing areas and the globe (land and sea surface) under four representative concentration pathway (RCP) scenarios of increasing greenhouse gas concentrations. Error bars represent standard deviation (SDs) in the climate model results. From Zhao et al. (2017), used under the terms of the Creative Commons Attribution 4.0 license.

2014; Voss et al., 2015). Behrenfeld and colleagues have modeled changes in oceanic productivity and found sustained declines over a decade in the ocean's phytoplankton biomass as a result of climate warming (McClanahan, Allison, and Cinner, 2015).

Climate Change—A Global Justice Issue

Climate change has the characteristics of a "global commons" problem. As well as being potentially disastrous for the Earth system and all its inhabitants, issues of equity and justice arise. GHG emissions by particular states contribute to the problem and affect all other states and people, while only the state responsible for producing emissions benefits directly from the associated productive and development outcomes. At the same time, no state has the capacity on its own to achieve effective mitigation and to ensure that anthropogenic emissions are contained within environmentally sustainable bounds. Equity and justice issues also arise in relation to power imbalances in international negotiations for action on climate change. In the absence of a mechanism that regulates the behavior of all states, individual states have an incentive to continue producing emissions at a rate that will ultimately ensure global environmental destruction. In doing so, they will contribute to the creation of intergenerational injustices.

Causal Responsibility, Future Development Rights, and the Unequal Distribution of Climate Change Harms

Together, the least-developed and developing economies have contributed, since the mid-eighteenth century, less than 25% of global cumulative CO_2 emissions (Lykkeboe and Johansen, 1975). While they now account for the bulk of growth in global emissions, their total contribution to annual emissions is still less than that of the developed world (Lykkeboe and Johansen, 1975). At the individual level, consumption by an average citizen in the UK accounts in just 2 months for as many GHG emissions as does a whole year's worth of consumption by an average citizen of a low-income country (UNDP, 2011). As shown in Table 1.1, China has recently overtaken the United States as the largest national emitter of CO_2, but its per capita contributions remain much lower than those from the United States (Friedlingstein et al., 2014).

> Even in an emergency one pawns the jewellery before selling the blankets. . . . Whatever justice may positively require, it does not permit that poor nations be told to sell their blankets in order that the rich nations keep their jewellery. (Shue, 1992)

In a seminal article published in 1992, the philosopher Henry Shue argues that, from an ethical perspective, people should not be asked to sacrifice their vital interests in survival in order to allow other people to continue living lives of relative luxury. This is particularly the case where the threat to some people's vital interests stems from a process that other people caused and from which their descendants continue to benefit. In the process of industrializing, developed countries largely exhausted the world's capacity to take up, redistribute, and sequester carbon—in other words, to absorb the GHGs released into the atmosphere as a result of human activities. By doing so, they have in effect denied other countries the opportunity to use their own shares of the Earth's absorption

Table 1.1 **Estimated CO_2 Emissions from Fossil Fuel Combustion and Cement Production for 2013 and 2014, with Growth Rates**

	2013				**2014**	
	Total ($GtCO_2$ $year^{-1}$)	**Per Capita (tCO_2 p^{-1})**	**Cumulative 1870–2013 ($GtCO_2$)**	**Growth 2012–13 (%)**	**Total ($GtCO_2$ $year^{-1}$)**	**Growth 2013–14 (%)**
World	36.1 (34.3–37.9)	5.0	1,430 (1,360–1,500)	2.3	37.0 (34.8–39.3)	2.5 (1.3–3.5)
China	10.0	7.2	161	4.2	10.4	4.5
US	5.2	16.4	370	2.9	5.2	−0.9
EU-28	3.5	6.8	328	−1.8	3.4	−1.1
India	2.4	1.9	44	5.1	2.5	4.9

$GtCO_2$ = Gigatonnes Carbon Dioxide per year; tCO_2 = total Carbon Dioxide per person.

Source: Friedlingstein et al. (2014), used under the terms of the Creative Commons Attribution 4.0 licence.

capacity to achieve comparable levels of fossil-fuel driven development. Access on equal terms to a vital global commons has thus been usurped. Shue argues it is, therefore, a matter of justice that poor nations are

> not asked to sacrifice . . . the pace or extent of their own economic development in order to help to prevent the climate change set in motion by the process of industrialization that has enriched others. (Shue, 1992)

In his view, developed states must bear the economic costs of climate change mitigation and adaptation, ensuring via technology transfers and financial relief that poor nations are able to continue developing and do not bear the opportunity costs of dealing with climate change (Shue, 1992).

The "vital interests" affected by climate change for members of developing states are not only in future development but also in surviving the harms that it has already caused and that will flow from it in the future. As I describe further later in this chapter and in Chapter 2, the impacts of climate change will be greatest in the world's poorest countries, both because of their geographic location and because they do not have the resources necessary to adapt to the effects of climate change. Many poor nations have densely populated and low-lying coastal regions that are highly vulnerable to sea level rises. A sea level rise of 1 meter would, for example, inundate 17% of Bangladesh and displace 18 million people in a country with few resources to cope with such enormous impacts and that in 2014 produced just 0.3% of global GHG emissions

(Harris, 2014). Other poor nations—primarily in Africa—are vulnerable to rising temperatures, drought, and desertification.

Butler (2003) argues that there are enough global resources, knowledge, and technology to provide an adequate standard of living for the world's current population, while containing climate change within limits that will ensure the survival of humanity and the planet. The United Nations Development Programme (UNDP) concurs, suggesting that vastly improved human development outcomes may be achieved via more equitable forms of global governance and distribution of resources, removing the need for continued pursuit of economic growth where this is dependent on fossil fuel consumption (UNDP, 2011). This is a topic I will return to in Chapter 3. As Butler admits, transitioning to a world in which all humanity's basic needs are met and further climate change is arrested is a utopian vision (Butler, 2003). Ensuring justice in how the world deals with climate change is difficult not only because of inequities in contributions to the harm, in who benefits from it, and in the risks it poses, but also because many low- and middle-income countries lack political and economic power and thus are at a disadvantage in negotiating the legal and policy frameworks that have been and are being adopted to respond to climate change.

Structural Injustice in Climate Change Negotiations

As we have seen, there is a chasm between those who historically contributed most to climate change and those most affected by it and their capacity to deal with it. It continues to be the case that those who are influencing decisions and future agendas on climate change are not accountable from the perspective of those most affected by the risks, and those most affected have limited ability to influence the decision-making process. While populous and rapidly developing states like China and India have negotiating leverage based on the large contributions to future emissions they will make in the absence of effective action on climate change, small island and less developed states like the Maldives, Haiti, and Bangladesh are highly vulnerable to climate change risks but have little leverage in international negotiations (Shue, 1992). This fact largely explains the world's continuing failure to reduce overall GHG emissions and to take the wide-ranging and radical action necessary to reliably avert the risks of climate change. Existing agreements do, however, acknowledge and attempt to redress—albeit to a limited degree—some of the inequities discussed here.

The founding international agreement on climate change, the 1992 United Nations Framework Convention on Climate Change (UNFCCC), now functions as the primary intergovernmental forum for negotiating global responses to climate change. At the outset, the UNFCCC recognized that developed and developing states have divergent responsibilities for climate change and capabilities to redress it, and it vested developed states with primary responsibility for combating climate change. The Convention specifies that:

> The Parties should protect the climate system for the benefit of present and future generations . . . on the basis of equity and in accordance with their common but differentiated responsibilities and respective capabilities. Accordingly, the developed country parties should take the lead in combating climate change and the adverse effects thereof (article 3(1)).

The 1997 Kyoto Protocol imposed binding emission reduction targets only on developed states, but the targets were extremely modest and, even so, Australia, Canada, and the United States subsequently refused to ratify the Protocol (Savaresi, 2016; Singer, 2002). The Protocol's first commitment period ended in 2012 and the negotiation of revised reduction targets was fraught, with Japan, New Zealand, and the Russian Federation refusing to adopt revised and more onerous targets (Savaresi, 2016). The Protocol allowed countries that had difficulty reaching their targets to engage in "emissions trading," purchasing emissions credits from countries that were able to reach their targets with emissions to spare, with the result that the overall impact on climate change was minimal (Singer, 2002).

The 2015 Paris Agreement, discussed in the Introduction, has been heralded as overcoming a long-standing impasse in negotiations on climate change action (Savaresi, 2016), but it has also been criticized as enshrining "many promises and little action" (Sanders, 2015). As noted earlier, the Agreement seeks to limit climate change within 2°C above pre-industrial levels, and adds an aspiration of keeping it within 1.5°C above these levels. The distinction between developed and developing country commitments has been replaced with a broader notion of differentiation that allows parties to devise their own commitments or "nationally determined contributions." Parties will be required to report regularly on their emissions and their implementation efforts from 2020.

Intergenerational Inequity

Another climate change injustice is that of intergenerational inequity. The emissions produced by past and current generations will have the greatest negative impact on future generations, as GHGs accumulate and the Earth system's capacity to absorb and adapt to them surpasses its limit. There is debate in philosophical circles about the status and strength of duties to future generations (Caney, 2008). Recognition of basic human dignity should, however, compel current generations to acknowledge a duty to prevent foreseeable and egregious harm to future generations where it is in their power to do so without compromising their own vital interests.

Health Inequity—What We Know So Far

I have been discussing the science of climate change and the inequities arising from differences in who has contributed the most to the problem, who has benefited the most from these climate change producing activities, and who is most at risk from the consequences of climate change. Among these consequences are major health-related threats. These are my focus in the final sections of this chapter, where I discuss how climate change interacts with other social determinants of health to create a multiplier effect, deepening and compounding global health inequities. Before turning to that discussion, I provide an overview of existing health inequities within and between countries. I also describe how recognition and understanding of the social-determinants-of-health inequities have evolved. I highlight the role of eco-social theory and eco-epidemiology in growing appreciation of the health-related impacts of climate change, but note the tendency to gloss over the connections between climate change and health inequities.

A Brief Status Update on Health Inequities within and between Countries

At one level, humanity is making progress. According to the World Health Organization (WHO), life expectancy at birth, globally, has been improving at a rate of more than 3 years per decade since 1950, with the exception of the 1990s when the human immunodeficiency virus (HIV) epidemic in Africa and the collapse of the Soviet Union led to an increase in mortality rates in those regions (Figure 1.5) (WHO, 2016e).

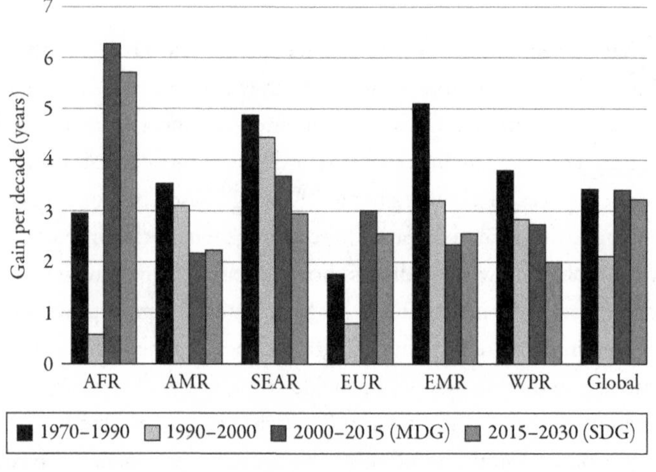

Figure 1.5 Regional and global life expectancy per decade, 1970–2030. MDG, Millennium Development Goal; SDG, Sustainable Development Goal; AFR, Africa; AMR, Americas; SEAR, Southeast Asia; EUR, Europe; EMR, Eastern Mediterranean; WPR, Western Pacific. From WHO World Health Statistics 2016. Used with permission.

Despite improvements in the averages, however, there are marked inequities between regions and countries. Inequities in the incidence of premature death and disease also exist within countries, and recent analysis by WHO suggests that health inequities within and between countries are growing (WHO, 2016b). It seems remarkable that, today, a man living in the East End of Glasgow, where this author is from, is at risk of dying 15 years earlier than a man living in the West End of Glasgow (Glasgow Centre for Population Health, 2014). In a prosperous country such as Australia, where I now live, the poorest 20% of the population can still expect to die younger—6 years on average—compared to the richest 20% of the population (Leigh, 2013). Those who are socially disadvantaged in terms of income levels, employment status, and education, and Aboriginal Australians—who are more likely to be disadvantaged in all these areas—are also at higher risk of depression, diabetes, heart disease, and cancers (Australian Institute for Health and Welfare, 2015). Similar inequities exist within countries everywhere.

Internationally, as Table 1.2 shows, the extremes of life expectancy span a remarkable 35 years—from 87 years for women in Japan to 51 years for women in Sierra Leone. One explanation for such vast divergences in life expectancy are similarly vast divergences in the incidence and impacts of disease. The burden of communicable disease has been reduced considerably throughout the wealthy world but continues to have a huge impact in poor countries. In 2015, for example, there were an estimated 214 million cases of malaria worldwide and 438,000 malaria deaths. More than two-thirds of these were children under 5 years of age, and around 90% of all cases were in Africa (WHO, 2015b). In the

Table 1.2 **Countries with the Highest and Lowest Life Expectancy at Birth (in Years), by Sex, 2015**

Male		**Female**	
Country	**Years**	**Country**	**Years**
HIGHEST		HIGHEST	
Switzerland	81.3	Japan	86.8
Iceland	81.2	Singapore	86.1
Australia	80.9	Spain	85.5
Sweden	80.7	Republic of Korea	85.5
Israel	80.6	France	85.4
Japan	80.5	Switzerland	85.3
Italy	80.5	Australia	84.8
Canada	80.2	Italy	84.8
Spain	80.1	Israel	84.3
Singapore	80.0	Iceland	84.1
LOWEST		LOWEST	
Lesotho	51.7	Chad	54.5
Chad	51.7	Cote d'lvoire	54.4
Central African Republic	50.9	Central African Republic	54.1
Angola	50.9	Angola	54.0
Sierra Leone	49.3	Sierra Leone	50.8

Source: WHO World Health Statistics (2016). Used with permission.

same year, almost half a million children under the age of 5 died from diarrhea-related illnesses, with rates of death highest in sub-Saharan Africa and South Asia (Troeger et al., 2017). Diseases transmitted by water, soil, and vectors, such as schistosomiasis, hookworm, malaria, and dengue fever, are generally much more common among people who suffer socioeconomic disadvantage such as poverty and poor housing conditions. A comparison of dengue cases in two cities on the US–Mexico border found a prevalence of 32% in Matamoros, Mexico, compared to only 4% in Brownsville, Texas. The increased infection rate in Matamoros was attributed to limited access to air conditioning, small housing-lot size, and limited use of insect repellents (Murray, Quam, and Wilder-Smith, 2013).

Vulnerability to disease is exacerbated by undernutrition. Some individuals, communities, and even entire nations do not have enough food. There is little food readily available, and many individuals are unable to afford what food there is. An inadequate intake of food, hence insufficient caloric intake, results in undernutrition. Undernutrition impedes physical development, causing stunting and wasting, and making it more likely that those who contract communicable diseases will die from them. Usually associated with a deficiency in micronutrients—as well as stunting and wasting, and impaired immunity—this can lead to impaired cognitive functioning, blindness, and poor reproductive outcomes (Bhutta, Salam, and Das, 2013; Gibson, 2011; Viteri and Gonzalez, 2002). Some progress has been made in reducing undernutrition globally when compared with the start of the century, but the improvements are very uneven (International Food Policy Research Institute, 2015), and the number of undernourished people has grown over the years since 2014 (FAO, 2017). In 2016, the number of undernourished people in the world increased to an estimated 815 million, compared to 777 million in 2015 and 775 million in 2014 (FAO, 2017). At the same time, 108 million people faced "crisis level food insecurity" or worse—a 35% increase over the previous year (World Food Program, 2017). "Crisis level food insecurity" involves "food consumption gaps with high or higher than usual acute malnutrition," or an inability to meet minimum food needs without depleting assets to a point where it will be impossible to meet these needs in the future (Food Security Information Network, 2017). Groups such as the Food and Agriculture Organization have expressed concern that we may be witnessing a new "upward trend" in undernourishment (FAO, 2017). This is also reflected in the prevalence of undernourishment assessed as a percentage of global population. While undernourished people constituted 14.7% of the world's population in 2000, this shrank to 10.8% in 2013 but has been almost stagnant or rising since then, reaching an estimated 11% in 2016 (FAO, 2017).

Undernutrition is not experienced equally, with people in low-income countries, pregnant women, and children most affected (Kyu et al., 2016). South East Asia, the Caribbean, and sub-Saharan Africa are all experiencing alarming rates of undernourishment—particularly Eastern Africa, where an estimated third of the population is undernourished (FAO, 2017). Largely because of the size of its population, the greatest number of undernourished people overall live in Asia (FAO, 2017). Maternal and child undernutrition is estimated to be the underlying cause of approximately 3.5 million deaths a year in low- and middle-income countries (Alessio, 2013; Bhutta and Salam, 2012; Black et al., 2008). In 2016, almost one in four children under the age of 5 were affected by stunting, the highest prevalence being in Africa (38%), followed by South East Asia (33%). Stunting is much more common among children born in rural areas or poor households, or to mothers with no education (WHO, 2016e). Wasting affects almost 52 million children under 5 years of age globally, with almost half living in South East Asia (FAO, 2017). Meanwhile, a third of all women of reproductive age suffer from iron deficiency anemia (FAO, 2017), which is also the leading cause of disability among children and adolescents globally (Wang et al., 2016). Among women the problem is more prevalent in developing countries (Mawani and Aziz Ali, 2016), and the vast majority of cases among children and adolescents are also concentrated in developing countries (Kyu et al., 2016).

Even when there is sufficient nutritious food available, health may be threatened through the ingestion of contaminated food. Microbial pathogens, chemicals, biotoxins, and parasites can contaminate food during the production and preparation processes (Hanson et al., 2012). The majority of foodborne illness is acute. It may cause renal disease, neurological disorders, and death (Lindsay, 1997). Diarrheal diseases, of which a considerable proportion are foodborne, are responsible for over 1 million deaths globally; as we saw earlier, almost half of these are children under 5 years of age (Troeger et al., 2017). While most of these deaths occur in low- and middle-income countries, foodborne disease also causes deaths in high-income countries, including an estimated 3,037 deaths annually in the United States (Ford et al., 2014; Hanson et al., 2012; Scallan et al., 2011).

At the same time as close to 1 billion people go hungry, there are almost 2 billion people who consume too many calories and who are overweight or obese (International Food Policy Research Institute, 2015). I discuss the nutrition transition to highly processed, calorie-rich diets and the role of an industrial food system in driving this transition in Chapter 2. The tendency to consume too many calories goes along with consuming too few micronutrients, which is contributing to burgeoning rates of noncommunicable diseases (NCDs). NCDs have emerged, relatively silently, as the leading cause of more than 60% of premature deaths worldwide. Over three-quarters are caused by cardiovascular diseases, cancer, diabetes, and chronic respiratory disease (Wang et al., 2016).

NCDs are inequitably distributed, varying by race, ethnicity, and income (International Food Policy Research Institute, 2015). For example, a study of obesity among adults in California found that almost a quarter were obese, but within different racial and ethnic groups the incidence of obesity varied, from a low of 9.8% among people with Asian backgrounds to a high of 36.1% among African Americans (Wang et al., 2017). Another example of variation in dietary consumption in the United States with significance for health equity involves attitudes toward piped drinking water. Negative attitudes play a role in encouraging increased consumption of bottled water and may also stimulate consumption of sugary soft drinks or juices rather than water (Pinard et al., 2013). Pinard and colleagues cite the research finding that distrust of publically supplied tap water and the belief that bottled water is safer than tap water in the United States was higher among young people, members of lower socioeconomic groups, and racial and ethnic minorities. These concerns were not irrational: there have been numerous cases of water quality problems in low-income and minority communities in the United States, as well as problems with contaminated piped water in schools (Pinard et al., 2013). While a global phenomenon, the rate of transition to calorie-rich and nutrient-poor diets is greatest in low- and middle-income countries where the healthcare burden of NCDs is growing rapidly (Popkin, Adair, and Ng, 2012; Reubi, Herrick, and Brown, 2016; Wang et al., 2016)—almost 80% of NCD deaths are in low- and middle-income countries (Reubi et al., 2016).

Air pollution—both ambient, or outdoor, and household, or indoor—is another major risk factor for NCDs. In 2016, up to 90% of people living in cities were exposed to fine particulate matter in concentrations exceeding WHO air quality guidelines, with exposure rates varying considerably by region (WHO, 2016a). Compared to the rest of the world, air pollutants are much higher in Africa, Asia, and the Middle East (WHO, 2016a), and 11 times more people die from indoor air pollution in low-income countries than elsewhere (UNDP, 2011).

Understanding the Social Determinants of Health Inequity

Understanding of the causal pathways to health inequities has come a long way. Social epidemiology has pried open the "prisoners of the proximate" shackles (McMichael, 1999), evolving from a focus on individuals' behavioral risk factors to consider the wider societal-level factors that contribute to health and health inequities—the "social determinants of health" (Commission on Social Determinants of Health, 2008). Epidemiologists now recognize that understanding people's behavior is impossible if this behavior is decoupled from its social context and, as we will see later in the chapter, from the natural environment (Berkman and Kawachi, 2000; Krieger, 1994, 2001; Marmot et al., 1991).

Good health requires sufficient income to feed and clothe oneself and one's family, and to access clean water and sanitation, decent shelter and living environments, healthcare, and education. Yet access to sufficient income and other material and social resources is not merely a matter of individual initiative and commitment. At the core of a political economy or materialist explanation of health inequities is the understanding that material circumstances, including the employment roles people occupy, their pay and other working conditions, and even the likelihood that they will obtain employment, are influenced by structural factors. This means that health and other inequities reflect political, economic, and social contexts (Scambler, 2007), including, for example, shifting global labor markets toward precariousness (Benach et al., 2014) and prevailing race and gender relations within societies. In all societies, rich and poor, the materialist hypothesis also suggests that social protection systems and services, such as education, health, transportation, and housing, are vitally important for health. Empirical studies from around the world provide compelling evidence in support of this, with a consistently graded relationship between socioeconomic position, living and working conditions, and health. Generally, the further down the social ladder one is, the greater the risk of poor health and premature death (Di Cesare et al., 2013; Labonté, Schrecker, and Gupta, 2005; Marmot et al., 1991).

Material deprivation helps explain the poor health of the impoverished, but it does not fully explain the social gradient in health. Bourdieu (1989) argues that individuals' life choices are constrained and their health influenced not only by the socioeconomic resources available to them but also by the very existence of a social hierarchy, their perception of the position they occupy within it, and their lived experience of that position. What this suggests is that while people need the basic material prerequisites for a decent life, they also need social affirmation. This theory is supported by numerous studies, but it was the landmark Whitehall cohort studies, involving British civil servants, that provided robust empirical evidence of a strong social gradient in health outcomes across economically secure occupational positions. What the studies found was that there was "a steep inverse association between social class, as assessed by grade of employment, and mortality from a wide range of diseases" (Marmot et al., 1991). Why, among people who are not poor in the usual sense of the word, should the risk of dying be intimately related to where they stand in the social hierarchy (Marmot, 2004)? The Whitehall studies demonstrated that, in addition to material security, a range of psychosocial factors are important for health and are contributors to health inequities, including social support, social capital, and social cohesion.

There is continuity between material and psychosocial resources through a third dimension, which has to do with power: the degree to which individuals, communities, and even nations are empowered to influence the decisions affecting the conditions in which they live (Popay et al., 2008). Health equity depends on inclusion and agency (Figure 1.6). This requires that individuals and groups have the capacity to make choices about how they will live from among a range of realistically available options, thus exercising meaningful control over their lives. It

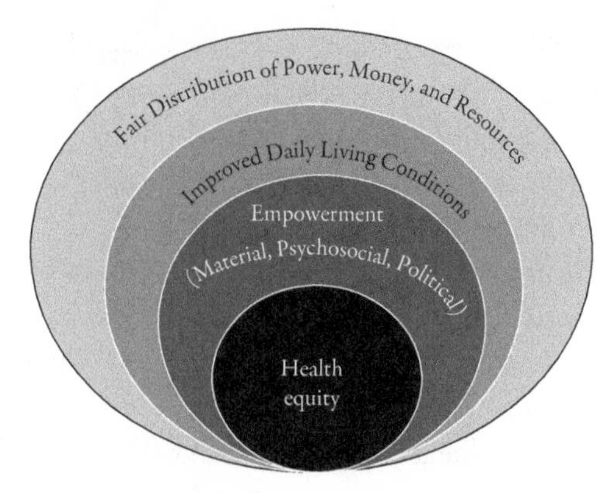

Figure 1.6 The social determinants of health equity conceptual model.

also requires that individuals and groups be able to represent their needs and interests strongly and effectively and, in so doing, to challenge and change current resource distributions.

Within any particular society, the amount and distribution of material and psychosocial resources, and the degree of individual and collective agency, are influenced by local, regional, and global factors. The role played by these factors, and the interplay between them, will vary according to the historical and cultural context. In the current era, international and global influences are increasingly dominant. We will see this in Chapter 2 in discussing the influence of the industrial food system on health outcomes, and it is obviously the case in relation to climate change.

Evolution of the Eco-Social Model of Health

Around the same time that knowledge about the social determinants of health was evolving, the field of environmental epidemiology was studying the effects on human populations and health outcomes of environmental exposures to materials such as air pollution, hazardous waste, metals, pesticides, and radiation. Environmental epidemiology tended, however, to take a somewhat reductive, mechanistic approach toward environmental health risks and associated inequities, and rarely considered the influence of social, economic, and political contexts or the interplay of local, regional, and international factors. Given that the environmental impacts of human activities have expanded exponentially, the focus needed to be enlarged. With climate change entering the mix of health hazards encountered and caused by humans, recognition has grown of the need to deal with environmental health problems on a new, global scale.

Appreciation of this global scale and of the range of influences on human health required a significant conceptual as well as methodological shift. In her pioneering work in the early 1990s, Krieger proposed the eco-social theory of well-being, health, and disease distribution. This advance in theoretical understanding argued that people embody their societal and ecological contexts throughout their life course—they "literally incorporate, biologically, the world around [them]" (Krieger, 1994, 2001). Shortly afterward, Susser and Susser proposed some methodological approaches which would enable the application of Krieger's eco-social theory. The field of eco-epidemiology emerged, with the proposition for an integrated approach to investigating disease and its prevention, meaning that consideration would be given to different levels and types of causal factors, different disease trajectories and outcomes, and notions of temporality (Susser and Susser, 1996). These theoretical and methodological perspectives are very relevant in the context of climate change and health. They recognize that understanding and promoting long-term good health requires an appreciation of the complex interactions between ecological, physical, and social environments, alongside more proximal and individualized risk factors.

Importantly, epidemiologists have started to acknowledge, research, and understand the role of climate change in the production of population health outcomes (Krieger, 2015; McMichael, 1994; McMichael and Beaglehole, 2000; Patz and Olson, 2006; Whitmee et al., 2015). Yet attention to ecology and to social systems and their interaction with physical systems remains somewhat missing in this mix. This partly explains why up to this point relatively little attention has been given to climate change and health inequities.

Climate Change—Exacerbating Existing Health Inequities

> [As a result of climate change] the rich will find their world to be more expensive, inconvenient, uncomfortable, disrupted, and colorless—in general, more unpleasant and unpredictable, perhaps greatly so. The poor will die. (Smith, 2008)

As we saw earlier in the chapter, climate change is causing more severe floods, droughts, storms, and heatwaves. It is associated with changes to hydrological systems, sea level rises, ocean acidification, species extinction, and increasing food and water insecurity, and it brings heightened risks to human societies and to human health

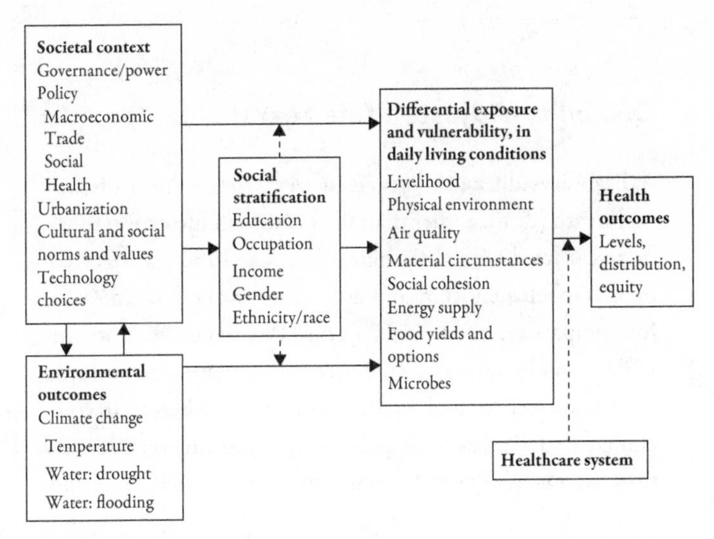

Figure 1.7 Direct and indirect pathways from climate change to health and health inequities. Source: Friel et al. (2008), used under the terms of the Creative Commons Attribution 4.0 licence.

and survival (Costello et al., 2009, McMichael A.J., 2009, Watts et al., 2018). Some of the health effects will be direct and immediate, while others typically will occur via more complex and indirect causal pathways (Figure 1.7).

In 2003, Campbell-Lendrum and colleagues estimated that climate change that had occurred over the decade from 1990 to 2000 was directly responsible for at least 5.5 million "disability-adjusted life years" in the year 2000 alone (Campbell-Lendrum, Corvalán, and Prüss-Ustün, 2003) (Box 1.3). This was a conservative estimate, because it did not include a range of likely effects of climate change that were considered too difficult to quantify, including the

Box 1.3

Disability-Adjusted Life Years

A disability-adjusted life year is either a year lost due to premature death (death earlier than the maximum life expectancy) or a year lived with a disability. The year 1990 was selected by Campbell-Lendrum and colleagues as the baseline year for comparison because in its Third Assessment Report the IPCC concluded that *the majority* of climate change since then had been caused by human activity, whereas it concluded that climate change since the mid-nineteenth century was *mainly* due to human activity (Campbell-Lendrum et al., 2003).

effects of changes in air pollution and aeroallergen levels; nor did it take into account the effects of climate change that had occurred prior to 1990 (Campbell-Lendrum et al., 2003).

A more recent, 2014 assessment by WHO projects that in 2030, climate change that has occurred since 1990 will be the cause of at least an additional 241,000 deaths in that year alone, in comparison to a world in which there had been no further climate change beyond 1990: 38,000 due to heat exposure in elderly people, 48,000 due to diarrhea, 60,000 due to malaria, and 95,000 due to childhood undernutrition (WHO, 2014). Further projections suggest that climate change will cause an additional 250,000 deaths *per year* between 2030 and 2050. Once again, these estimates are conservative because they exclude a range of known effects that are difficult to quantify (WHO, 2014).

While nowhere and no one will be untouched by climate change, the health risks and impacts will not be experienced equally, between regions and nations or among groups within individual countries (WHO 2014, 10). Climate change will add to pre-existing infectious and noncommunicable disease burdens that already weigh heavily in poor countries with under-resourced healthcare systems. Failure to reduce background rates of disease and premature mortality will ensure that the multiplier effects of climate change greatly exacerbate existing global health inequities.

Of the 250,000 additional deaths each year that climate change is likely to cause between 2030 and 2050, the majority will be in developing countries (Patz et al., 2007; WHO, 2014). The loss of healthy life-years in low-income African countries is predicted to be 500 times that in Europe. Increases in the global disease burden that have already occurred as a result of climate change have overwhelmingly affected developing countries. WHO estimates that 99% of the increased disease burden has been concentrated in these countries (Patz et al., 2007). Projections for annual climate change–attributable mortality resulting from undernutrition, malaria, diarrheal disease, dengue fever, and heat suggest that sub-Saharan Africa is likely to be the region most severely affected up until 2030, and South Asia is likely to be most affected by 2050 (WHO, 2014) (Figures 1.8).

It is notable that the greatest health risks posed by climate change are and will continue to be experienced by those contributing least to its underlying causes, and in regions and

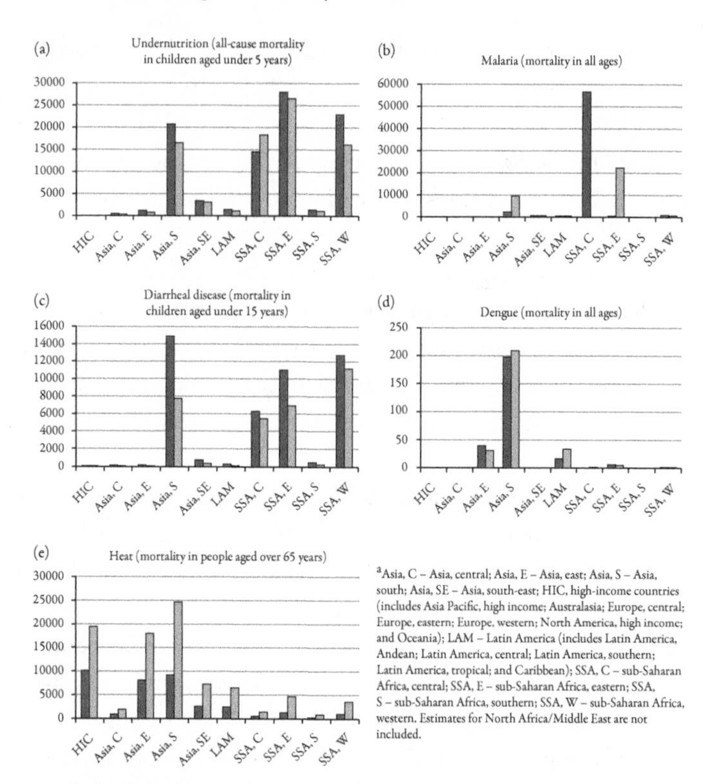

Figure 1.8 Regional distribution of annual climate change–attributable mortality projected in 2030 (blue bars) and 2050 (orange bars) from (a) undernutrition, (b) malaria, (c) diarrheal disease, (d) dengue, and (e) heat. Source: WHO (2014), with permission to display from the World Health Organization.

nations already experiencing high disease burdens and inequities (Figure 1.9).

Within countries, adverse health outcomes will be greatest among poor people living and working in urban and coastal areas,

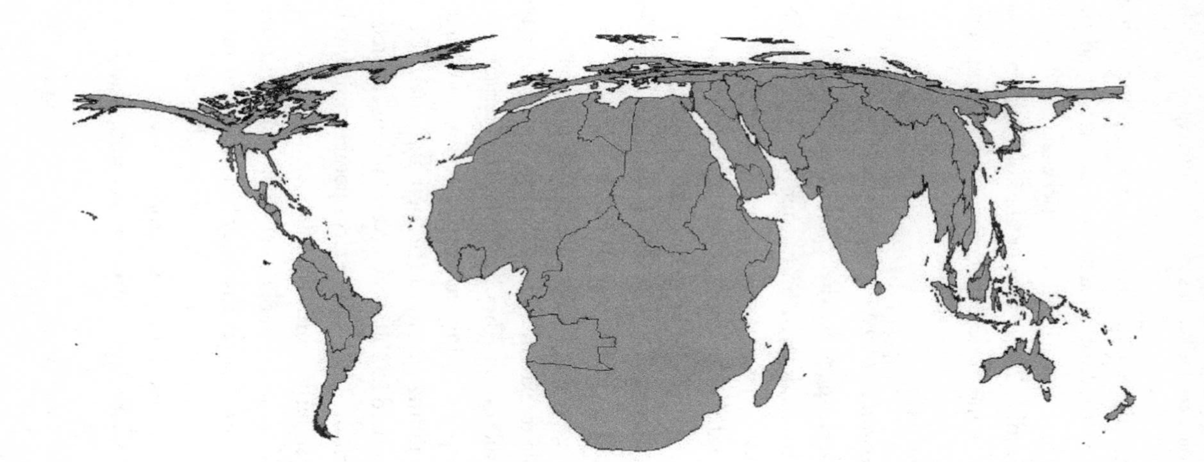

Figure 1.9 Deaths attributable to anthropogenic climate change between 1970 and 2000, density-equaling cartogram. Source: Patz et al. (2007), used under the terms of the Creative Commons Attribution 4.0 licence.

elderly people, children, traditional societies, and subsistence farmers, who have the least capacity and resources to adapt to climate change or to mitigate its effects (Friel et al., 2008; Kjellstrom, Holmer, and Lemke, 2009; Patz et al., 2007). Even in wealthy countries, existing inequities will deepen and be compounded, as described next.

Health Inequities and Extreme Weather Events

Weather disasters such as the cyclones that struck coastal populations of Bangladesh, Myanmar, and Vietnam in 2016 injure and kill. The hurricanes that frequently impinge on the southern United States and the Caribbean cause property damage, injury, deaths, and distress. Floods in northern Kenya killed and injured people directly, and also caused outbreaks of Rift Valley fever, affecting both livestock and humans (Friel et al., 2011).

Research suggests that disadvantaged communities are particularly vulnerable to extreme weather events, especially a series of events in close succession. They have limited material resources to plan for and respond to impending disasters or to engage in recovery and rebuilding afterward (Bennett and Friel, 2014). This was clearly illustrated in 2005 when New Orleans suffered the onslaught of Hurricane Katrina. The city's poorest residents lived in its lowest-lying suburbs, which were flooded and destroyed. Most of the emergency evacuation plans relied on residents driving out of the city to places of safety that they had located in advance for themselves. Yet many residents of the worst-affected suburbs did not own cars, and they did not have the money to arrange transport out of the city and alternative accommodation (Fussell,

Sastry, and Vanlandingham, 2010). Most of the people who died in the flooding and destruction caused by the storm came from these disadvantaged populations. These were also the groups that suffered most in the aftermath, as a result of the loss of their homes and possessions, and as a consequence of the broader effects such as damage to infrastructure, displacement, loss of livelihood, and halting economic recovery. It was also lower-income groups, and in particular children and the elderly, who were at increased risk of developing severe mental health symptoms compared with their peers in higher-income groups (Bennett and Friel, 2014; McLaughlin et al., 2009).

Rising Sea Levels and Health Inequities

Coastal inundation, more extensive flooding, increasingly severe storm surges, and damage to coastal infrastructure, including roads, housing, and sanitation systems, all pose direct climate change–related risks that will deepen existing global health inequities. We are witnessing rapid growth in the numbers of people living in low-lying coastal regions (below 10 meters in elevation)—the majority of this growth is in developing countries (Dodman et al., 2013; Neumann et al., 2015). In the year 2000, 10% of the world's population lived in low-lying coastal zones and 83% of these were in less developed countries (Neumann et al., 2015). Modeling by Neumann and colleagues (2015) suggests that population density in these zones will increase further and that the highest level of exposure to the effects of sea level rises, flooding, and storm surges will be in developing countries in Asia. Entire low-lying small island states that are mostly very poor face the prospect of losing all inhabitable land. Along with it, they stand to lose their culture, histories, and, potentially, their very identity as independent peoples.

Indirect risks to health from sea level rise include salination of freshwater supplies, transmission of infectious diseases, loss of productive farm land, and changes in breeding habitats for coastal-dwelling mosquitoes. Other indirect health risks include the mental health consequences of property loss, break-up of communities, and displacement and migration, each with implications for physical health as well (Durkalec et al., 2015; McMichael, 2016; McMichael, Barnett, and McMichael, 2012).

Heat Stress and Health Inequities

Some research suggests that heatwaves are responsible for more fatalities than other climate hazards such as floods and hurricanes (Reckien et al., 2017; Satterthwaite et al., 2007).

There are various pathways from heat stress to population-level health and social impacts, as shown in Figure 1.10. Heat kills people primarily by causing heart attacks, strokes, and respiratory failure. Heat-related health risks vary among social groups, with older people, infants, women, and people with pre-existing medical conditions at increased risk. This is also true of people who suffer socioeconomic disadvantage, including in the spheres of education, employment, ethnicity, and home ownership. People in these groups are particularly vulnerable to heat stress due to the factors described next.

Most of the world's poorest people live at low latitudes and are therefore exposed to more frequent daily temperature extremes than richer populations (Luke et al., 2016). Climate change will

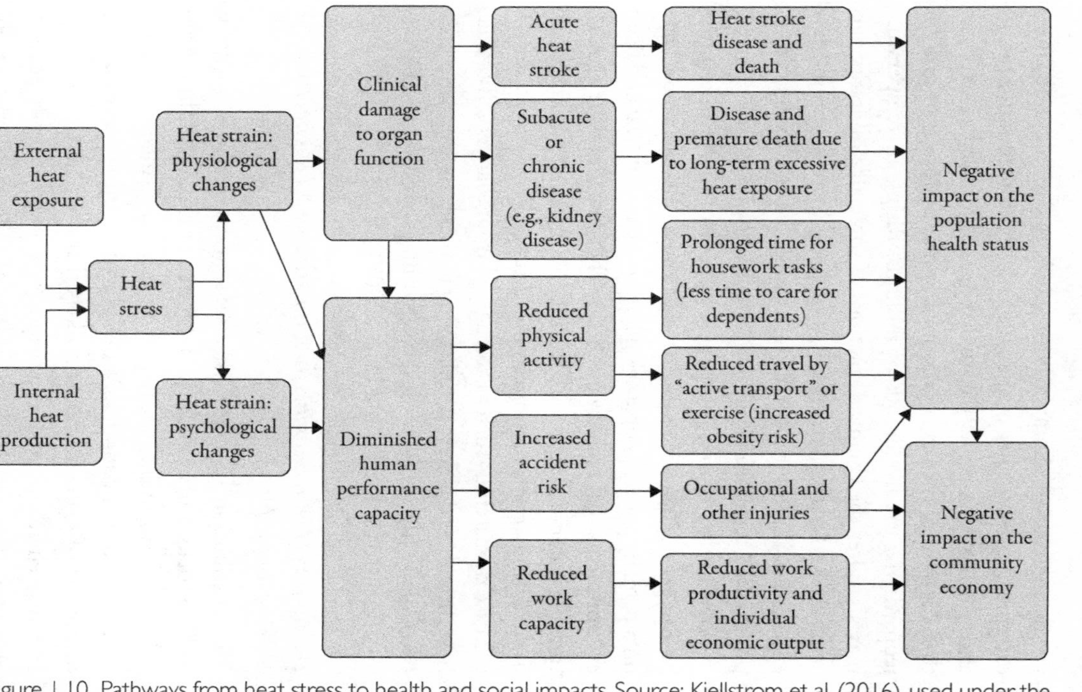

Figure 1.10 Pathways from heat stress to health and social impacts. Source: Kjellstrom et al. (2016), used under the terms of the Creative Commons Attribution 4.0 license.

greatly increase the number of people who live in conditions so extreme that the ability of the human body to maintain heat balance during physical activity will be compromised. This means that, for parts of the year, outdoor labor may no longer be possible, jeopardizing the livelihoods of millions. Rising air temperatures may also alter the chemical reactions of some air pollutants, such as ozone and particulate matter, and could increase ground-level exposure to air pollution or allergens while potentially aggravating pre-existing respiratory and cardiovascular diseases.

While relatively well-off people can afford to live in insulated buildings and have access to air conditioning and clean, cold water on tap, many poorer people are unable to escape the heat and live in environments that amplify its effects. They may not have access to running water or electricity. They may live in dwellings made from makeshift materials that offer no protection from the heat or, in the case of materials such as corrugated iron, that actually radiate heat. Infants and elderly people living in these environments are at particularly high risk for heat-related illness and death (Bennett and Friel, 2014; Kovats and Hajat, 2008).

Pathways to Health Inequities from Environments Hospitable to Vector-Borne Diseases

The distribution of many infectious disease vectors, including rodents, ticks, and mosquitoes, is limited by the availability of suitable habitats, and they are highly sensitive to changes in temperature, humidity, and rainfall. While global warming may lead to the extinction of many species, it is likely that it will at the same time enhance environmental conditions hospitable to disease hosts or vectors, facilitating the expansion of their range and seasonal endurance (Patz et al., 2000; Short, Caminade, and Thomas, 2017). This would be disastrous for poor people, particularly those living in low- and middle-income countries who have limited access to preventive healthcare. For many of these people, protective measures such as bed nets, insect repellents, and vaccinations represent luxuries (Bennett and Friel, 2014). Moreover, as we saw earlier, communicable diseases already have a vastly disproportionate impact on people living in developing countries. Poor health and deaths from infectious disease have terrible ripple effects, creating burdens of grief and loss, reducing the productivity and health of workers, and entrenching already existing economic disadvantage (Bennett and McMichael, 2010). The effects of diseases such as malaria, meningitis, and tuberculosis can also be devastating for children's physical and cognitive development, reducing their chances for a healthy and productive adult life (Bennett and Friel, 2014).

Climate Change–Related Inequities in Availability and Affordability of Nutritious Food

As discussed earlier, climate change affects food production and availability through changes in average temperatures and rainfall patterns, by placing stress on water quality and availability, creating conditions hospitable to pests and disease, reducing biodiversity, and contributing to species extinction. In addition, changes in hydrological systems and an increase in severe weather events such as storms and flooding are contributing to increased soil erosion and degradation, and crop damage (IPCC, 2007, 189–190; IPCC, 2014a, 6, 13). Already, climate change is exacerbating existing inequities in the availability and affordability of nutritious foods between and within countries. In the future, climate pressures on food production will affect what is available for local domestic consumption as well as what is available in terms of global trade. Various pathways from some key climate impacts to food insecurity and malnutrition are shown in Figure 1.11.

These impacts are and will continue to be greatest in regions where the majority of the world's poorest and most vulnerable people live (Nelson et al., 2009, 4). The negative effects of climate change on agricultural yields are occurring primarily in countries in the tropics and subtropics, where already high levels of food insecurity exist (Baldos and Hertel, 2014; Barnett, 2011; IPCC, 2014a; World Bank, 2010b). A meta-analysis projects that, because of climate change, mean crop yields may decline across Africa and South Asia by 8% by the 2050s (Wheeler and von Braun, 2013), whereas the IPCC prediction for average global yield declines in the same period is in the region of 5% (IPCC, 2014a). As discussed earlier, droughts and long-term drying conditions are also becoming more frequent and severe in many already impoverished regions. The result is hunger, starvation, displacement, and misery. A third of the world, including some of its poorest people, depend primarily on agriculture for their livelihood (Grist, 2015). As a result of both climate change and the increasing reach of the industrial food system, which I discuss in Chapter 2, many farming jobs have already been lost, and suicide rates among farmers have increased (Berry et al., 2011; IPCC, 2014a; McMichael, 2017).

We saw earlier that climate change is likely to have huge impacts on fisheries and aquaculture, reducing marine biodiversity and limiting the productivity of aquaculture. Modeling also suggests there will be a large-scale redistribution of global catch

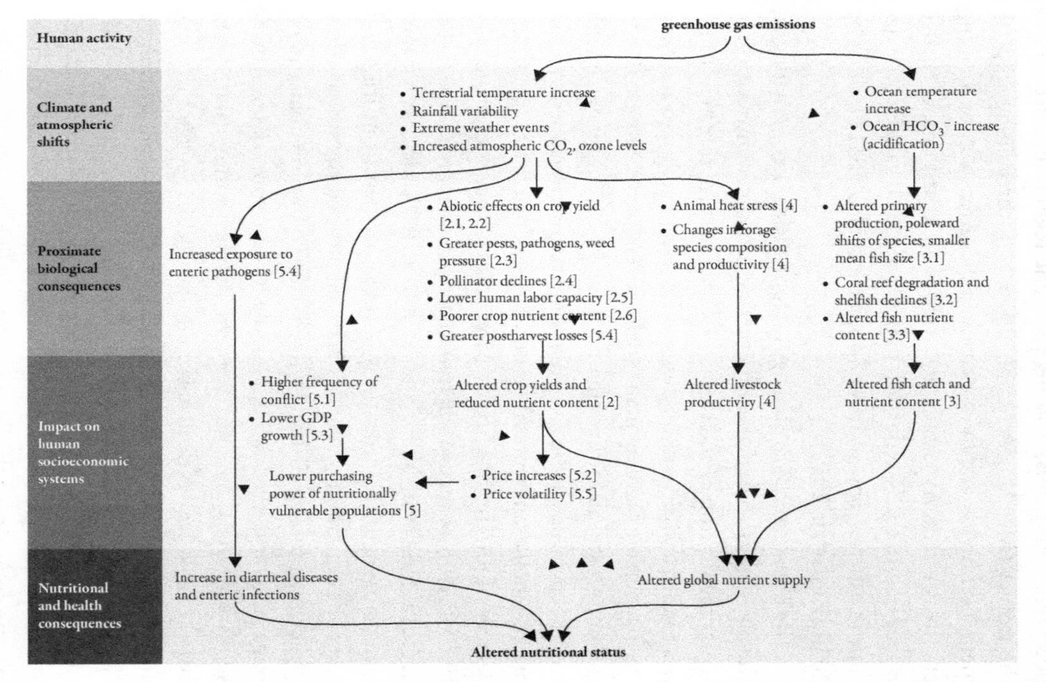

Figure 1.11 Pathways for impacts of climate change on food systems, food security, and undernutrition. GDP, gross domestic product. Source: Myers et al. (2017), used under the terms of the Creative Commons Attribution 4.0 licence.

potential, with increases in high-latitude regions and declines in the tropics (McClanahan et al., 2015, 84), threatening employment, livelihoods, and food security in the latter. A study focusing on the western African region estimated that climate change would lead to a 21% reduction in the landed fish value, a 50% decline in fisheries jobs, and a total annual loss of US $311 million by 2050 (Lam et al., 2012).

The impact of climate change on food yields is exacerbating an already fragile social and health situation in many countries, mainly across the subtropics and tropics. Climate change threatens progress toward a world without hunger. It creates barriers to attaining the Sustainable Development Goals (SDGs), especially Goal 2, to end hunger and all forms of malnutrition by 2030. Some estimates have been made of the potential impacts of climate change on dietary intake and mortality outcomes, albeit with significant degrees of uncertainty. Springmann and colleagues estimate that by 2050, climate change will lead to per-person reductions of 3.2% in global food availability, compared to a 10.3% increase without climate change, and reductions of 4.0% per person in fruit and vegetable consumption and 0.7% in red meat consumption. They suggest that food availability will be reduced by more than the average amount in the low- and middle-income countries of Africa, South East Asia, and the Western Pacific (Springmann et al., 2016) (Table 1.3). Nelson and colleagues (2009) suggest that climate change will result in a 20% increase in child malnutrition by 2050, relative to a world without climate change. Reduced fruit and vegetable consumption is considered the main risk factor for death in high-income countries and in the low- and middle-income countries of the Western Pacific, Europe, and the Eastern

Table 1.3 Global and Regional Food Availability and Consumption of Fruits, Vegetables, and Red Meat, 2010 and 2050, for Reference Scenario without Climate Change and Mean (SD) of Main Climate Change Scenarios

Baseline in 2010		Model Scenario, 2050	
		Reference Scenario (without Climate Change)	Climate Scenarios, Mean (SD)
FRUIT AND VEGETABLE CONSUMPTION (G/PERSON/DAY)			
Global	342.2	378.0	363.1 (2.7)
High-income countries	375.9	397.7	382.3 (2.7)
LMICs of Africa	196.5	242.3	233.2 (1.8)
LMICs of the Americas	324.1	362.3	348.7 (1.8)
LMICs of the Eastern Mediterranean	332.4	340.8	327.7 (2.1)
LMICs of Europe	314.0	366.0	352 6 (3.1)
LMICs of South East Asia	215.3	321.8	3.7.3 (2.6)
LMICs of the Western Pacific	539.0	602.1	579.2 (4.3)
RED MEAT CONSUMPTION (G/PERSON/DAY)			
Global	62.2	66.5	66.0 (0.1)
High-income countries	135.8	133.9	132.8 (0.2)
LMICs of Africa	18.2	36.4	36.0 (0.1)
LMICs of the Americas	89.8	99.3	98.4 (0.1)

(*continued*)

Table 1.3 **Continued**

Baseline in 2010		**Model Scenario, 2050**	
		Reference Scenario (without Climate Change)	**Climate Scenarios, Mean (SD)**
LMICs of the Eastern Mediterranean	19.4	37.1	36.9 (0.1)
LMICs of Europe	70.8	78.8	78.2 (0.1)
LMICs of South East Asia	9.1	14.1	14.0 (0.0)
LMICs of the Western Pacific	101.7	126.0	125.3 (0.1)
TOTAL KCAL AVAILABILITY (KCAL/PERSON/DAY)			
Global	2817.5	3106.9	3008.3 (10.8)
High-income countries	3414.3	3433.6	3372.5 (10.8)
LMICs of Africa	2417.6	2878.7	2756.5 (12.9)
LMICs of the Americas	2886.4	3051.5	2979.0 (10.8)
LMICs of the Eastern Mediterranean	2661.4	2932.2	2855.4 (14.8)
LMICs of Europe	3035.4	3256.1	3198.8 (16.0)
LMICs of South East Asia	2406.5	2856.9	2740.6 (13.4)
LMICs of the Western Pacific	3016.9	3512.8	3401.5 (15.9)

LMICs, low and middle-income countries; SD, standard deviation.

Source: Springmann et al. (2016), used under the terms of the Creative Commons Attribution 4.0 licence.

Mediterranean. Undernourishment is the primary risk factor in the low- and middle-income countries of Africa and South East Asia. Increases in the number of people who are overweight or obese are the main risk factors in the low- and middle-income countries of the Americas (Springmann et al., 2016).

Climate change will also affect dietary intake indirectly via food prices, which will reflect the impact of climate change on production yields and also on the cost of fuel and animal feed. The IPPC considers it very likely that changes in temperature and precipitation will contribute to increased international food prices by 2050, with estimated price increases ranging from 3% to 84% (IPCC, 2014a). These estimates do not take into account the potential benefits for crop yields of increased CO_2 in the atmosphere. The IPCC notes, however, that the bias from omitting positive CO_2 effects may be counterbalanced in the studies on which its assessment was based, because negative effects of increased O_3 levels and increases in pest and weed damage associated with climate change were also omitted (IPCC 2014a, 512). Lobell and colleagues (2011) estimate that food prices will increase by 6% by 2050 as a result of the impact of changes in temperature and precipitation on food supply, taking into account the beneficial yield effects of increased CO_2.

Different crops will have different levels of vulnerability to price pressures resulting from climate change, for example, because of the kind of inputs on which they rely, as well as because of their direct vulnerability to changes in temperature and precipitation. Nelson and colleagues found climate change is likely to have the greatest impact on wheat prices compared to other important crops such as maize, rice, and soybeans. Meat prices will also be affected, largely because of the rising cost of animal feed (Nelson et al., 2009).

While globally most agriculture relies on rainfall, 20% of all agricultural land is cultivated using irrigation, and this 20% supplies 40% of the world's food (FAO, 2011, 23). Water scarcity is already a critical concern in parts of the world (Fedoroff et al., 2010). Water shortages as a consequence of reduced rainfall may lead to increased demands for irrigation water, putting pressure on water prices, while high demand for land in productive zones with reliable rainfall or other water supplies may increase the price of land in these areas (Fischer et al., 2007; Hanjra and Qureshi, 2010). The extent to which these factors will have an impact on food prices is likely to be considerable, but the specifics are largely unknown.

Concluding Remarks

This chapter has focused on the injustices associated with global environmental change and in particular climate change. By highlighting the facts of climate change, the historical and continuing inequities in emissions produced by developed as compared to developing countries, asymmetries in the negotiating power of states, and inequities associated with the failure to account for the interests of future generations, the chapter demonstrates the multiplicity of injustices that may be perpetuated by climate change.

The chapter also surveyed some important contemporary ideas concerning the social explanations for health inequities between and within regions and countries. Climate change is contributing to health problems throughout the world and is doing so in an uneven manner, compounding existing inequities. The chapter suggests that underpinning many of the pathways to climate change–related health inequities are inequities in the social conditions in which people live and the systems that contribute to these social conditions. These social determinants will also be affected by climate change, through impacts such as disruption to livelihoods, reduced material resources, and loss of a sense of control over one's life. These will be greatest and felt most acutely among socially vulnerable groups (AP-HealthGAEN, 2011; Friel et al., 2008, Marmot et al., 2008). Importantly, the IPCC has now acknowledged this interrelationship between climate change, health outcomes, and other social factors such as daily living conditions and access to social services (IPCC, 2014a; Woodward et al., 2014).

The next chapter examines the interplay between all of these issues and draws attention to the consumptagenic system that is driving both climate change and health inequity.

2 IT'S A CONSUMPTAGENIC WORLD

PRODUCING CLIMATE CHANGE, EXACERBATING HEALTH INEQUITIES

Introduction

Chapter 1 highlighted how the effects of climate change interact with the social determinants of health to exacerbate health inequities within and between countries. In Chapter 2, the focus is on an underlying cause of both climate change and health inequity. This is the dominance of what I am calling a "consumptagenic system"—a network of policies, processes, and modes of understanding and governance that fuels unhealthy, inequitable, and environmentally destructive production and consumption. The consumptagenic system poses a fundamental challenge confronting humanity right now.

In the first part of this chapter, I describe the evolution of a globalized market-based economic system characterized by fossil fuel dependence and the relentless pursuit of growth. I discuss the development of the consumptagenic system as a component part of this economic system and emphasize its addiction to conspicuous overconsumption. The consumptagenic system drives both climate change and health inequity, and climate change further exacerbates health inequity. This means that we cannot redress health inequity without tackling climate change, and we cannot tackle climate change without constraining the consumptagenic system. From a positive perspective, if we confront the consumptagenic system head on, and the economic system that enables it, we will be able to reduce the harms of climate change at the same time we move toward more equitable health outcomes. The survival of our planet and the well-being of its people depends on our ability to reign in consumption and to change its character.

To make all of this concrete, the chapter examines the roles played by an industrial food system and urbanization within the consumptagenic system. Why focus on the industrial food system and urbanization? Because both are central cogs in the consumptagenic system, fueling forms of consumption that contribute hugely to global warming. Both are at the same time centrally implicated in the production and deepening of health inequities. In Chapter 1, I noted how climate change is aggravating health inequities by undermining food security. In this chapter, on the industrial food system, I show how this system is also producing other food-related health inequities, at the same time driving climate change. My discussion of urbanization similarly demonstrates not only the role that urbanization plays in driving climate change but how climate change will have an impact on cities to exacerbate health inequities.

Fossil-Fueled Growth—The Backbone of Consumptagenic Systems

Planetary and human health are under threat from a consumptagenic system. This system is itself enabled by and a product of a globalized market-based economic system predicated on growth and shaped by actors and governance structures that mostly disregard health, social equity, and environmental consequences and limits (Whitmee et al., 2015). Even when health, equity, and environmental concerns are taken into account, this tends to occur in a manner that defers to the logic and supremacy of capitalist market thinking, which assumes that the continued expansion of production and consumption is a fundamental and primary good.

Seismic global transformations that have taken place over the past few centuries, including the industrialization processes of the eighteenth century and the economic globalization that has occurred throughout the twentieth century, have served to entrench capitalist market thinking (Polanyi and MacIver, 1944). Its influence has extended into spheres where it was previously uncommon—for example, through the privatization of education provision, health services, water management, and traditional knowledge, in the latter case via intellectual property regimes (Larking, 2018). Although often defended on the basis that unfettered markets are the most efficient distributors of goods and services, ensuring that goods are produced and services provided as economically as possible; that creativity and constant improvement are incentivized; and that the needs of consumers are met, the problems associated with the reach of capitalist market thinking and how it has influenced policy development are legion. These include acknowledged—but mostly unmitigated—market failures, such as monopoly influence and the creation of externalities.

Monopoly influence is evident, among many other places, in global food and agricultural markets, where just a handful of firms now dominate global production and sales (Clapp, Desmarias, and Margulis, 2015; Ghosh, 2010; Larking, 2017; Thow et al., 2015). More on this topic is provided later in the chapter. The problem of externalities is fundamental for dealing with climate change and other environmental stresses. Externalities arise when there is a failure, or it is not feasible, to factor either negative or positive consequences of a production process into how products are priced, usually because the producers do not bear the costs or benefits of these consequences. A striking example is pollution, which may be hard to trace back to particular producers, making it difficult to ensure they pay the costs of remediation (Larking, 2018). Moreover, understanding of the disastrous long-term damage caused by some pollutants, including greenhouse gases (GHGs), is only starting to be understood. Even given the contemporary and growing appreciation of these consequences, there is a lack of political will to ensure that they are dealt with and the costs borne by the producers. The difficulty of effectively allocating costs and of ensuring that they are paid for by those responsible arises not only because much of the damage was done many years ago, but also because of the character of the environment and Earth systems as a global commons, as we saw in Chapter 1.

Perhaps the most fundamental issue with the current dominance of capitalist market thinking is its dependence on growth, and particularly on forms of economic growth that deplete natural resources and are highly polluting. Throughout capitalism's history, economic growth has been accompanied by environmental damage (Jabobs and Mazzucato, 2016). Why is this so? The engine of capitalism—production—is "fueled" by fossil fuels, including coal, oil, and gas. Over millennia, large amounts of carbon from decayed plants and animals have been deposited into the earth in the form of fossil fuel. In the endeavor to expand the production of commodities in order to accumulate capital, or make more money, technological developments advanced in such a way that they required the burning of increasing volumes of fossil fuels to power machinery and transportation within and for the various industries. This, of course, meant that fossil fuels were mined from the earth, used to produce commodities or to transport them, and put back into the atmosphere in the form of GHGs, particularly CO_2. This has been happening at a volume and rate faster than the environment can absorb them. As a result, and as warned by scholars such as Daly (1974), we have the current climate crisis: the life support services of nature have been depleted, with all of the flow on effects for human health and survival described earlier in the book.

In the consumptagenic system, it is taken for granted that economic growth, conceived in the form of annual increases in gross national product (GNP), is desirable (Simms and Johnson, 2010). Within a globalized market economy, growth has become a standard-bearer for, and marker of, progress (McMichael, 2017).

Consumptagenic Societies

The fossil fuel–dependent, market-based economic system described in Box 2.1 relies on people wanting to consume the

Box 2.1

Dependence of a Globalized Market-Based Economic System on Fossil Fuels

As the dominance and reach of the market-based economic system grew, it became dependent on a constant supply of fossil fuels and the energy sector was structured around fossil fuels. In total, 80% of the world's energy still comes from oil, gas, and coal (International Energy Agency, 2016).

Australia and Japan have the dubious title of being fossil fuel leaders of the developed countries, with 90% of their energy coming from fossil fuels. The powerhouse BRICS economies (Brazil, Russia, India, China, and South Africa) are among the largest emitters of greenhouse gases in the world, owing to their enormous production and consumption of fossil fuels, consuming more energy from fossil fuels than any other source (Downie and Williams, 2018). Coal is the largest source of energy demand in China, India, and South Africa, whereas oil and gas are the main sources in Russia and Brazil. As large energy consumers, China and India have an interest in reducing their dependence on imported fossil fuels, whereas Russia and Brazil, as large producers of oil and gas, have a very different interest, namely in increasing exports and demanding higher prices (Downie, 2015).

goods that the system produces, otherwise economies will not grow and wealth will not be accumulated. A major concern, from an environmental perspective, is that this creates excessive consumption, which in turn creates a sea of plastic, the collapse of habitats and species, and too many GHG emissions for the atmosphere to absorb without destabilization of the Earth system (Ellen MacArthur Foundation, 2016; Maniates, 2010; Motesharrei et al., 2016).

There are many economic, political, and sociological processes within the consumptagenic system that influence what, how much, and why we consume. Lipovetsky (2005) has described three phases of consumption, the first phase beginning in the 1880s and lasting until after the Second World War (Box 2.2). As technology developed, along with economic growth, it brought transportation, telecommunications, and mass marketing. Although consumer demand for goods grew, it was relatively limited, as the

Box 2.2

Acceleration in Production and Sale of Consumer Goods

"In 1932, in the United States, there were only 740 vacuum cleaners, 1580 ironing machines and 180 electrical kitchen ranges for 10,000 people. In France, in the same year, there were 120 vacuum cleaners and only 8 electrical kitchen ranges for 10,000 people" (Esposti, 2012).

incomes of the majority were insufficient to make many discretionary or lifestyle-oriented purchases. By the 1950s, there was a significant expansion in purchasing power among consumers worldwide. Increasingly available consumer goods became widely affordable, facilitated by greater access to consumer credit.

The Spread of Market-Based Economies

The modern consumptagenic system really emerged toward the end of the 1970s and into the early '80s. Scholars like Esposti (2012) and Lipovetsky (2011) refer to this as the "hyperconsumption phase." This phase is associated with an increasing number of capitalist market-based economies, generating an unprecedented increase in the desire and ability to consume.

The ripple effect from developed to developing countries of embracing market-based economic policies was greatly enabled by two major global policy developments. The first was the creation of structural adjustment programs by the World Bank and International Monetary Fund (IMF), which gave loans to developing countries to ease their balance-of-payment problems on older debts, on condition of national policy reform. These programs required countries to open their economies to market forces by liberalizing trade, investment, and the financial sector; deregulating and privatizing nationalized industries; removing regulatory controls on private-sector activity; and devaluing their currency and tightening monetary policy (Labonté et al., 2009). In essence, debtor countries were required to become part of the capitalist market-based economic system.

International trade agreements were also important drivers of the adoption of market-based policies and of consumption.

Described in more detail in the food system section later in this chapter, World Trade Organization (WTO) member countries were required to open their markets by reducing tariffs and non-tariff barriers to trade and investment. WTO rules promoted the global integration of national markets and provided a more favorable operating environment for the private sector, for example, by protecting intellectual property (Chase-Dunn et al., 2000). Also, the focus of trade and investment policy increasingly shifted toward vertical integration of global supply chains, enabled by greater liberalization of foreign direct investment and trade in goods and services (Baldwin, 2006).

Trade and investment liberalization opened countries to trade and investment in processing, manufacturing, and retail, and to advertising by international companies. These companies set about creating new mass markets in products that not only were unavailable previously but for which no markets had existed. Thirty years ago, for example, it was unheard of to buy bottled water where tap water was clean and abundant. By 2016, one million plastic bottles were being sold around the world every minute—most of them in China and the Asia Pacific region—creating an environmental crisis some campaigners predict will be as serious as climate change. Major drinks brands produce the greatest numbers of plastic bottles, with Coca-Cola producing more than 100 billion disposable plastic bottles every year (Ellen MacArthur Foundation, 2016). International trade and investment agreements across a whole range of sectors have assisted the treadmill of production of more goods and services. They have facilitated the creation of systems that stimulate and drive demand with little, if any, consideration of the lasting environmental costs.

International trade is now a major driver of global carbon emissions, with large quantities of emissions embodied in exports from emerging economies (Liu et al., 2015; Lykkeboe and Johansen, 1975). Up to a quarter (between 20 and 25%) of CO_2 emissions globally can be traced to the production of internationally traded products (Afionis et al., 2017; Hausfather, 2017). The standard method of attributing GHG emissions to particular countries—and the method that continues to be used in the United Nations Framework Convention on Climate Change (UNFCC)—is based on emissions generated at the point of production (Afionis et al., 2017). Yet given the increasing significance of international trade, and the role of consumption in stimulating production, there should perhaps be a greater focus on the demand side of the production equation. A "consumption-based" approach to tracking GHG emissions attributes emissions embodied in goods and services to the final consumer, and to the country in which those goods and services are consumed, rather than to the country in which they are produced. On this basis, particular countries can be characterized as "net importers" or "net exporters" of GHG emissions, depending on whether consumption-based emissions outstrip production-based emissions, or vice versa (Afionis et al., 2017) (Figure 2.1). China is by far the largest net exporter of emissions globally, outstripping Russia—the next largest exporter—by a multiple of five (Hausfather, 2017). The United States is the largest net importer of GHG emissions, accounting for double the amount of net import-based emissions as Japan, the world's second-largest net importer (Hausfather, 2017).

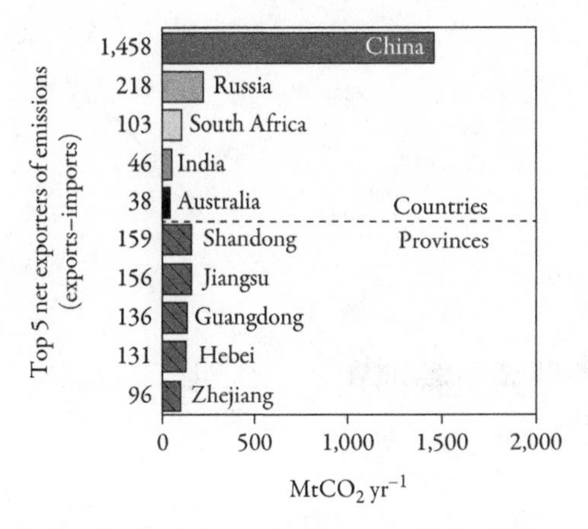

Figure 2.1 Top five net exporters of emissions. Note: Emissions embodied in trade. Top ten regions (including top five countries and top five Chinese cities/provinces) by emissions embodied in net trade, shown in absolute numbers. Data are for 2007. Source: Liu et al. (2015), used under the terms of the Creative Commons Attribution 4.0 licence.

Pricing Regimes and the Promotion of Cheap "Luxury" Goods

Beyond the arena of international trade and investment, a range of regulatory regimes exerts influence at a more household level, stimulating demand for goods and, hence, consumption, and also producing norms that legitimize consumer demand and conspicuous consumption. These include marketing and price incentives that promote the production and consumption of lifestyle and disposable goods while imposing barriers to the enjoyment of less environmentally damaging goods and services. The economist Mark Perry has shown that, between 1996 and 2016, the cost of necessities in the United States exceeded the overall inflation rate

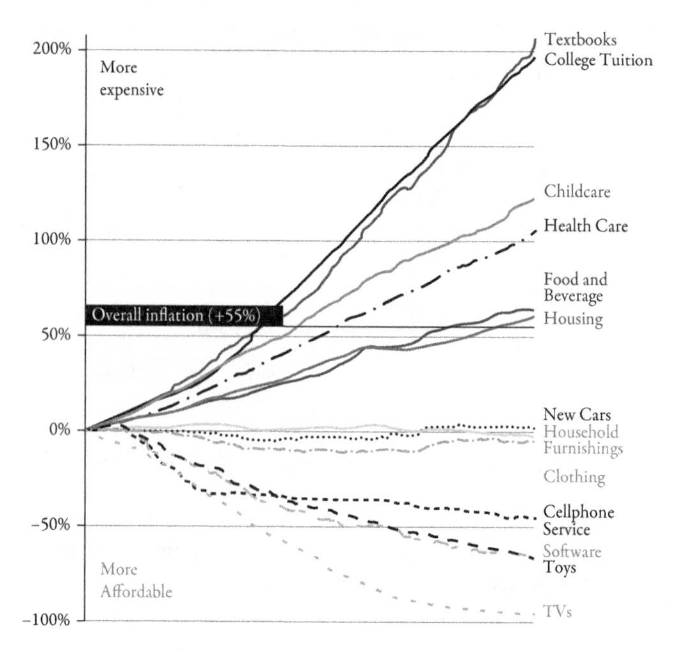

Figure 2.2 Price changes 1996–2016 of selected consumer goods and services. Source: Ingraham (2016), used under the terms of the Creative Commons Attribution 4.0 licence.

and in some cases more than doubled it (Ingraham, 2016) (Figure 2.2). During this period, the inflation rate was around 55%, while the cost of food and housing increased by close to 60%, healthcare and childcare costs more than doubled, and the price of textbooks and higher education nearly tripled. On the other hand, the cost of consumables like mobile phones, toys, software, and televisions fell drastically over the same period. In the words of journalist Christopher Ingraham, "For many Americans . . . that shiny new flat-screen TV is now more within reach financially than it's ever been. But it has become harder to afford the house to put it in, food to eat in front of it, or the medical care to ensure you'll outlive its extended warranty" (Ingraham, 2016).

Consumptagenic Food Systems and Their Cultivation of Consumptagenic Societies

We live in a food-rich world, but as we saw in Chapter 1, it is also a world in which malnutrition caused by the inability to access sufficient quantities of food or by excess consumption of poor nutritional-quality or contaminated foods is a major cause of illness and death. It affects all countries, and nearly one in three people worldwide (International Food Policy Research Institute, 2015). Why are some people hungry while, at the same time, others have plenty of nutritious food, and yet others eat food in large quantities but derive little nutritional benefit from it? The underlying causes of malnutrition and how it impacts different populations reflect a complex mix of factors, one of which is the operation of an industrial food system that has, particularly since the middle of the twentieth century, increasingly influenced what and how food is produced, distributed, and marketed within states and internationally. At the same time, the industrial food system is a major contributor to climate change and is, in turn, affected by it. Chapter 1 highlighted the fact that climate change is having an increasingly significant impact on the global incidence and distribution of malnutrition. If unchecked, the effects could be catastrophic, exacerbating rates of malnutrition and bringing death and disease to millions of already struggling individuals and households. My focus here is on how excess consumption is stimulated by an industrial food system that is contributing to climate change and health inequities.

Evolution and Dominance of the Industrial Food System

A *food system* can be defined neatly as the path that food travels from field or farm to fork (Anand et al., 2015). As well as the growing, harvesting, processing, packaging, transporting, marketing, consuming, and disposing of food, the concept of a food system includes all of the inputs needed and outputs generated along the way (Oxfam Australia, 2018). It also includes the mix of social, political, economic, and environmental factors that influence and are influenced by food supply pathways (Pinard et al., 2013). Food systems contribute to or undermine nutrition and food security through their influence on the amount of food available for consumption, its safety, nutritional quality, price, physical accessibility, and cultural acceptability (Friel and Ford, 2015; Pinstrup-Andersen, 2013).

Historically, the most common food systems were based on local production for local markets (Anand et al., 2015). Many of these localized systems have now been displaced by the modern industrial food system. The industrial food system dominates all other food systems, operating across sectors, including production, distribution, advertising, retail, and consumption. It also operates globally. International supply chains are made up of national and transnational food commodity producers, while multinational distributors and retailers include major supermarket chains and food service companies. Although globally dominant, the industrial food system sits alongside, and competes with, alternative commercial, national, and local food systems that involve producer co-ops, community-supported agriculture, artisanal farms, and civic agriculture based on household and community gardens (James and Friel, 2015; McCullough et al., 2008).

The development and eventual dominance of the industrial food system can be traced as a strand within the broader historical evolution of the consumptagenic system described here earlier. After WWII, the market-based economic system's principles of specialization, economies of scale, and accumulation of capital were applied to agriculture and to food processing, distribution, and marketing. Food shortages in war-torn Europe, plus perceptions that agricultural sectors in the developing world were underperforming, stimulated governments to intervene to promote production (Lang, Barling, and Caraher, 2009). The 1933 Agricultural Adjustment Act in the United States, and Europe's Common Agricultural Policy, established in 1957, set the scene for massive commodity production programs (Dimitri et al., 2005; Grant, 1997). The 1960s and 1970s in particular were decades of intensive state-sponsored investment in agricultural research and development. The World Bank and development organizations also invested heavily in agricultural programs in developing countries. The result was a so-called Green Revolution, during which wheat, maize, and rice yields soared. During this time, intensive meat production in the United States also began to be promoted (Glaeser, 2010; McMichael, 1994).

Many of the technologies developed during that period have transformed agricultural practices and yields globally, producing greater intensification of production but requiring recurring and costly inputs, often consuming vast amounts of water. Chemical fertilizers, pesticides, herbicides, and new seed varieties, along with irrigation, were the stimulants of yield increases (Neff et al., 2011). The Green Revolution also spurred massive land-clearing for crop production and permanent pasture. Crop production and permanent pasture now cover approximately 38% of the Earth's ice-free land surface, including around half of all former temperate deciduous forests and savannas, and almost three-quarters of the world's grasslands (Clark and York, 2005). As in many industrialized countries, most food crops eaten in the United States are produced on large land tracts planted in monocultures. Animals are raised separately from crops in large confinement facilities, eating specially formulated feeds.

Accelerating yields in the United States and Europe produced large surpluses that enabled these regions to become huge exporters of staples such as wheat, corn, rice, and soybeans, as well as cash crops, particularly sugar cane and other sources of sugar (Anand et al., 2015). This was a taste of what was to come in the 1980s and 1990s in terms of the manufacture and export of highly processed and nutrient-empty foods. In Latin America and Asia, agricultural growth rates exceeded population growth, but per capita food output in Africa declined. Globally, malnutrition remained high (Hawkes et al., 2012).

The 1980s saw a marked shift, globally, toward market-oriented economic policies, including agricultural and food policy and greater liberalization of trade and investment policy. In the previous section, I discussed the influence of structural adjustment programs implemented by the World Bank and the IMF which forced developing countries to open their domestic markets to international trade and investment. I also noted the influence of the WTO and its requirement that member countries reduce tariff and other barriers to trade and investment. This has had wide-ranging implications for the global food trade. Following the Uruguay round of multilateral trade talks in 1994, WTO member countries were required to open their agri-food markets by reducing tariffs and non-tariff barriers to imports, reducing subsidies for exports, and reducing domestic agricultural support.

These policy changes enabled the rapid expansion of the industrial food system, along with the structural inequities that are internal to it. The WTO trade liberalization provisions have had an asymmetrical impact, in effect serving the interests of wealthy states and corporations at the cost of developing states and small-scale producers and retailers (De Schutter, 2009; Larking, 2017, 2018). For example, the provisions relating to export subsidies allowed developed countries to retain their subsidy schemes, though reducing subsidies over time, but developing countries, most of which did not have subsidy schemes in place during the relevant baseline period from 1986 to 1990, were prevented from introducing them (De Schutter, 2009; Larking, 2018).

WTO rules promote the global integration of national food markets, including through harmonization of food safety and quality regulations, and provide a favorable operating environment for the private sector. Increasingly, investment liberalization, which aims to attract investment in processing, manufacturing, retail, and advertising by international companies, has become an important feature of trade policy (Friel et al., 2013; Hawkes, 2005). Consequently, control of the global food supply chain has shifted to large agri-food processors and transnational manufacturing, retail, and food service companies (Baines, 2014; Havinga, van Waarden, and Casey, 2015; Lang, 2003; McMichael P., 2009), thus affecting the availability of locally produced foods and the livelihoods of local producers (Clapp et al., 2015; Ghosh, 2010; Larking, 2017; Thow et al., 2015). Almost 90% of the global grain trade is controlled by just four agribusinesses. In the arena of processing and manufacturing there is also increasing corporate concentration. Ten food processors and manufacturers control almost a third of the global market. Nestle, Pepsi-Co, and Kraft are the top three most profitable firms that manufacture agricultural products into food products. Walmart, Carrefour, and Tesco are the most profitable food retailers that sell these foods to the consumer (EcoNexus, 2013).

While this is happening globally, the big concern is in emerging economies. With their large populations and growing wealth, these countries are particularly attractive investment targets for commodity-producing companies (Baker and Friel, 2016; Friel et al., 2013; Patel, 2012; Schram et al., 2015; Wilkinson, 2009). Greater penetration and control of the market increases the buying and selling power of transnational food companies, allowing them to dictate terms of trade, set buying and retail prices, preferentially promote particular goods, and influence eating habits through the products they choose to manufacture, sell, and market (Baker and Friel, 2014, 2016; Lang and Heasman, 2004; Monteiro et al., 2011; Moodie et al., 2013).

On the face of it, reductions in barriers to trade and investment should increase consumer food choices and improve supply of nutritious foods for net food importing countries. However, these developments have led to a global decline in the availability and diversity of nutritious food—undermining food security and exacerbating rates of undernourishment—and to the excess availability, affordability, and acceptability of highly processed, nutrient-poor foods (Box 2.3). This stimulates overconsumption of these foods and is driving the ballooning rates of noncommunicable diseases noted in Chapter 1 (Friel et al., 2013; Hawkes, 2005; Popkin,

Box 2.3

Two Examples of What Can Happen to Domestic Food Supply as a Result of Trade and Investment Liberalization

In Fiji and Samoa, trade and investment liberalization is associated with a decrease in the availability of starchy staple foods and the increased availability of nontraditional refined cereals and processed foods. The opening of markets has allowed an influx of cheap food products and the dumping of unhealthy goods such as fatty meats. Overall, liberalization has undermined the competitiveness of domestic agriculture and contributed to import dependency and increased consumption of unhealthy foods (Thow et al., 2011). The low cost of some imported foods reflects a mix of policy factors in major food and agricultural exporters such as the United States where, for example, farmers have been provided with income support that has allowed them to sell grain at prices lower than their production costs (Clark et al., 2012).

In Mexico, the investment provisions in the North American Free Trade Agreement (NAFTA) enabled US agribusiness investment throughout the food supply chain, creating challenges for local agricultural production (Clark et al., 2012). In cases in which escalating production costs led to increased prices for primary commodities, food-processing companies responded by substituting lower-priced but less nutritious ingredients, thereby influencing the nutritional quality of foods available in the marketplace (Hawkes et al., 2012).

2017). As discussed there, the burden of malnutrition weighs particularly heavily on low- and middle-income countries and on low socioeconomic groups within all countries (Brinkman et al., 2010; Hawkes, 2006; McCorriston et al., 2013; Popkin, 2017; Stuckler and Nestle, 2012).

Price Drivers of Inequity within the Industrial Food System

As noted earlier in the chapter, price is a key driver of a consumptagenic system, promoting certain forms of consumption at the cost of others and stimulating excess consumption. Within the industrial food system, nutritious food becomes less and less affordable while unhealthy foods are cheap, widely available, and marketed aggressively. The result is that wealthy people are able to eat healthy foods whereas poorer people find this more difficult to do.

Many factors contribute to global and domestic food prices (Bradbear and Friel, 2013) (Figure 2.3). Price drivers at the

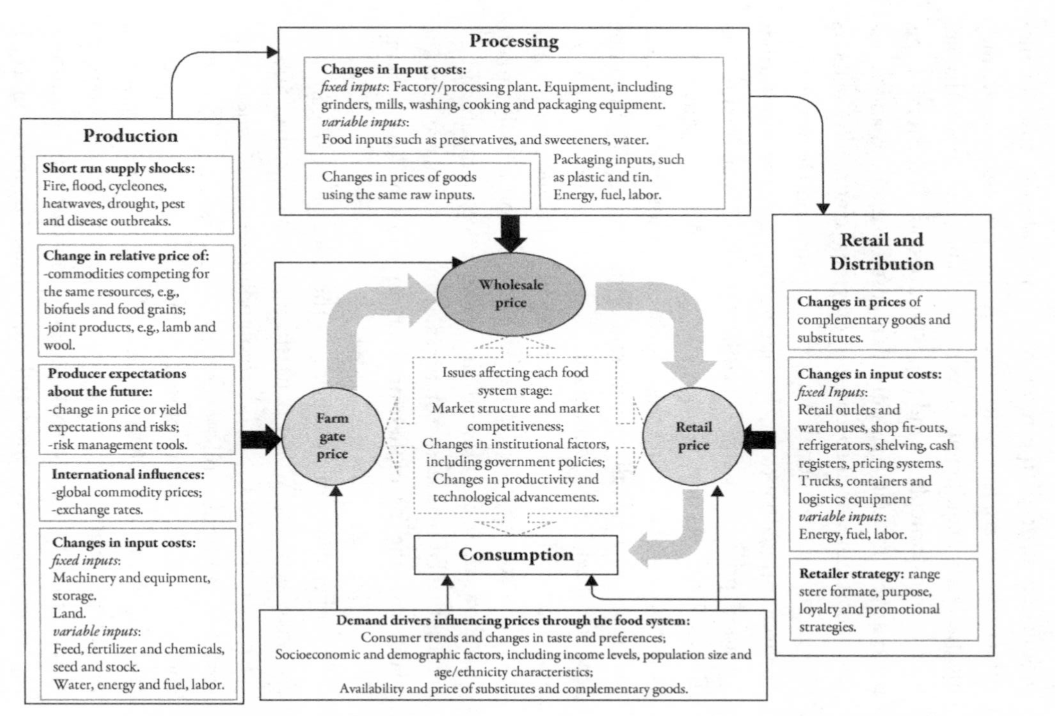

Figure 2.3 Supply and demand factors driving price through stages of the food system. Source: Bradbear and Friel (2013), used under the terms of the Creative Commons Attribution 4.0 licence.

production stage of the food system include the costs of machinery, water, seeds, and fertilizer. At the processing stage, they include factory equipment, preservatives, and other inputs that are used to transform plants and animals into food. At the retail stage, price drivers include transportation costs, marketing, and transaction costs. Financial speculation on global commodity markets also impacts international food prices (Baffes and Dennis, 2013).

Concentrated control of the supply chain has major implications for food prices and their volatility (Kalkuhl, von Braun, and Torero, 2016). As noted earlier, four agribusinesses control almost 90% of the global grain trade. These same businesses have a virtual monopoly globally on the sale of farming inputs, including seed, fertilizer, and agrochemicals, and at the same time monopolize markets for agricultural commodities, buying primary produce, and storing it for later release and sale on global markets. This allows these businesses to dictate the prices of inputs and the returns that farmers receive on the sale of agricultural products and to influence the cost of these products on global markets (Larking, 2018; Wilson and Edwards, 2008).

Another driver of food prices is the price of oil (Figure 2.4). An analysis by the World Bank suggests that more than 50% of price increases in five food commodities—maize, wheat, rice, soybeans, and palm oil—over two 8-year periods, 1997–2004 and 2005–2012, was accounted for by increases in the price of crude oil (Baffes and Dennis, 2013). High energy prices raise food prices because of the increased cost of inputs such as fertilizers and chemicals, and the increased cost of transporting both agricultural inputs and produce (Food and Agriculture Organization 2011;

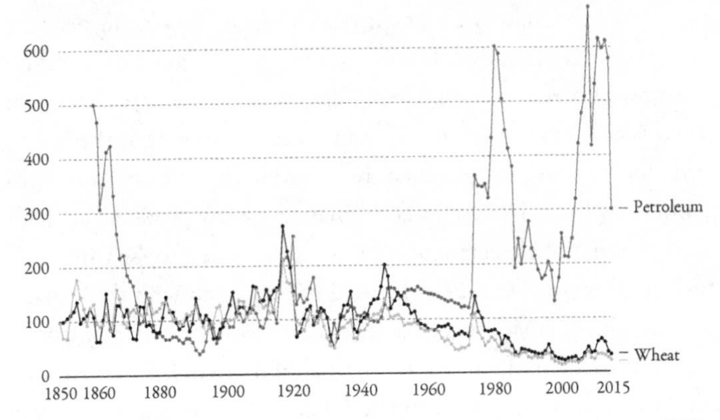

Figure 2.4 Commodity price index of cereal crops and petroleum, 1850–2015. Source: Roser and Ritchie (2018), used under the terms of the Creative Commons Attribution 4.0 licence.

Piesse and Thirtle, 2009). Increases in energy prices also influence the demand side of the equation. High fuel prices create new markets for agricultural crops that can be used for biofuels. Land and water resources traditionally used for food crops are increasingly being diverted to biofuel crops (Cottrill et al., 2007; Pelletier et al., 2011).

Rapidly increasing demand for meat and animal products globally, including in low- and middle-income countries, is also having a significant impact on global food prices. It means that more land is being used for feed crop production and less cereals are being produced for human consumption, placing price pressure on the limited stocks that are available for consumption (FAO, 2017; Popkin, 2017).

We saw in Chapter 1 that climate change is also placing pressure on food prices because of its impact on production yields and on the cost of fuel and animal feed. The impact of climate change on food production intersects with many other aspects of the industrial food system to influence the affordability and availability of unhealthy food. In 2008, a combination of low production yields and international speculation on food prices led to "soaring food prices" globally (World Bank, 2012). Combined with the impacts of the global financial crisis and accompanying fall in incomes and rise in unemployment, escalating food prices led to starvation and undernutrition. An estimated 48.6 million people who might otherwise have escaped poverty were unable to do so because of the rising cost of food (World Bank, 2012). In 2011, another food price spike sent a shiver across the world. International food prices jumped for the second time in 3 years, leading to a repeat of the 2008 food price crisis (World Bank, 2012). While things have improved somewhat since then, the Food and Agriculture Organization's food price index suggests that by the end of 2017 international food prices had almost doubled since the start of the century (Baffes and Dennis, 2013; FAO, 2017) (Figure 2.5)

The price of food in international markets, although of critical importance for food availability and affordability, does not always correlate directly with prices in national markets, which are influenced by a range of factors including exchange rates and government policy. Comparing 2007 with 2013, in low and middle income countries average real domestic prices increased by approximately 19% for rice and wheat and 29% for maize (Dawe et al., 2015). These are the three main cereals consumed by humans and they currently provide nearly two-thirds of global dietary

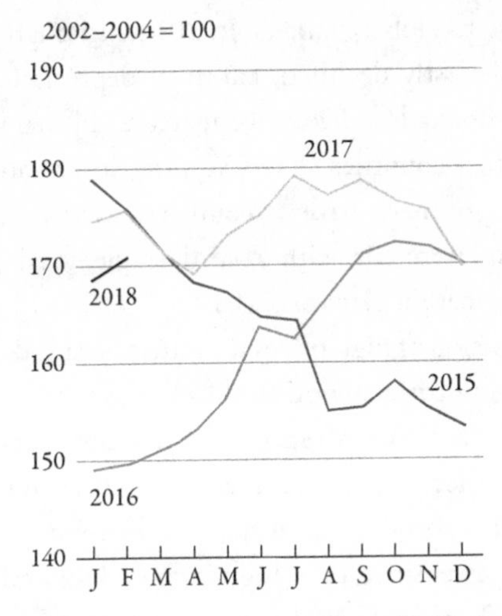

Figure 2.5 Food and Agriculture Organization (FAO) food price index, 2017. Source: FAO (2017), used under the terms of the Creative Commons Attribution 4.0 licence.

energy intake (Jones and Ejeta, 2016). Cereal prices are particularly significant for poor people because cereals constitute around 50% of their dietary energy supply, and higher prices may lead to decreased spending on more nutritious foods including green leafy vegetables (Dawe et al., 2015).

During the period 2007–2013, Dawes and colleagues (2015) found "striking" regional and national variations between price rises. For example, the three largest low- and middle- income countries—China, India, and Indonesia—contained the extent of domestic cereal price rises by banning exports, ensuring the availability and relative affordability of domestic supply, but adding price pressure to global markets. In 2016, although international cereal prices mostly declined, currency depreciations and local production shortfalls led to price increases in many low-income and food-deficit countries (Food Security Information Network, 2017). Some southern African countries, as well as Nigeria and South Sudan, were hit with record staple food prices (Food Security Information Network, 2017).

A combination of high prices for nutritious food, relatively low prices for less nutritious food, and lack of financial resources can lead to significant changes in purchasing and dietary behaviors. High food prices have the greatest impact on low-income households that spend a greater share of their income on food and are more price sensitive than high-income households (Nicholls et al., 2011; Thow et al., 2010). In many low- and middle-income countries, a high proportion of households now spend almost half of their income on food. This is in stark contrast to some high-income countries where, on average, households spend less than 10% of their income on food. In 2015, this was the case in Australia, Canada, Ireland, Switzerland, the UK, and the United States (Roser and Ritchie, 2018) (Figure 2.6). These figures do not mean that food is more expensive in low-income countries than in high-income countries. While the cost of food is often lower in low-income countries, average disposable incomes are also much lower, reducing the pool of money available for purchasing food. Even in high-income countries, however, there are inequities in spending power. In the United States in 2015, the poorest 20% of households spent on average 15.4% of their income on food, compared with an average of 11.2% spent by the wealthiest 20% of households (Bureau of Labor Statistics, 2015).

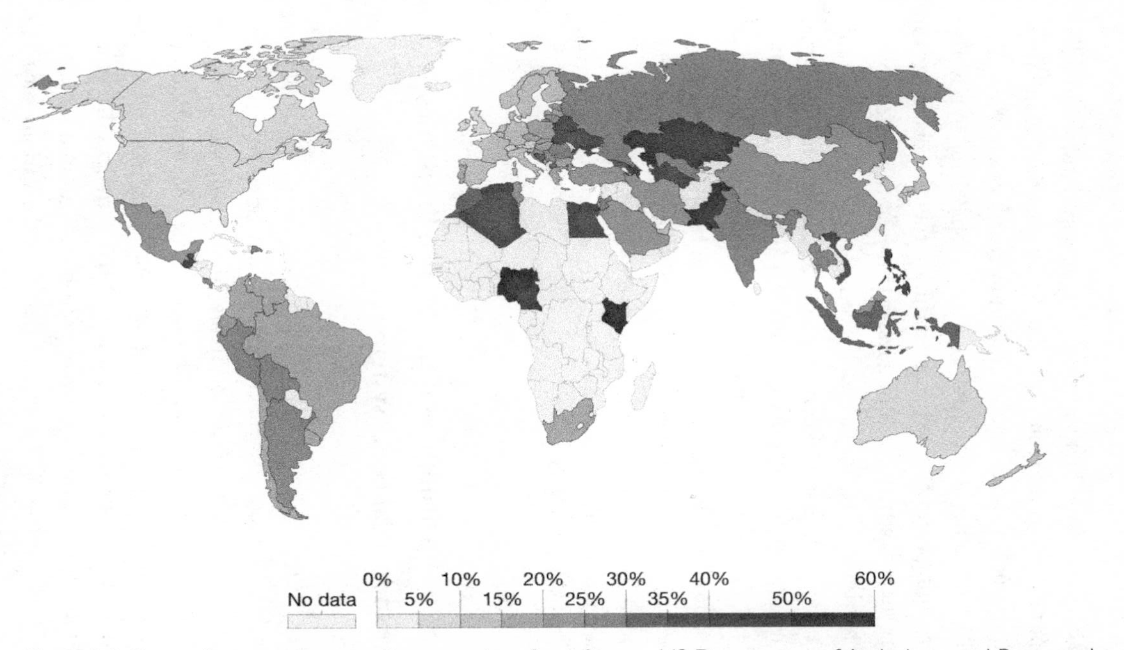

Figure 2.6 Share of consumer expenditure spent on food. Source: US Department of Agriculture and Roser and Ritchie (2018), used under the terms of the Creative Commons Attribution 4.0 licence.

When household finances are under pressure, food spending is commonly seen as more flexible than other essential expenditure demands, such as housing and education costs, which often take priority over food (Dowler, 2008). Many people living in poverty in developing countries were forced during the financial crisis to change their diet by substituting less expensive but nutrient-poor food for their usual food, and by consuming fewer meals and selling livestock assets (Economic and Social Commission for Asia and the Pacific [ESCAP], 2009). An analysis of 136 studies from 162 different countries found that increases in the domestic price of foods result in greater reductions in food consumption in poor countries—thus widening the inequities between countries. Within all countries, the model suggests that poorer households are the most adversely affected by increases in food prices (Green et al., 2013) (Table 2.1).

Emergent Environmental and Health Properties of the Industrial Food System

Eating Earth

The scale of the food-system changes just described represents an enormous assault on global ecosystems. All stages of the industrial food system, from production through to the decomposition of food waste in landfills, produce greenhouse gases and therefore contribute to climate change. Garnett (2016) estimates that globally, food systems generally are responsible for around a quarter of all GHG emissions. Figure 2.7 describes the various ways in

Table 2.1 **Mean Percentage Change in Food Demand for 1% Increase in Food Price by Country Wealth Category**

Food Group	Household Wealth Category	
	Lowest Income (*n* = 178)	**Highest Income (*n* = 177)**
Fruit and vegetables	−0.86 (−0.97 to −0.76)	−0.73 (−0.84 to −0.62)
Meat	−0.95 (−1.07 to −0.82)	−0.81 (−0.93 to −0.69)
Fish	−1.01 (−1.17 to −0.84)	−0.87 (−1.04 to −0.70)
Dairy	−0.92 (−1.08 to −0.78)	−0.79 (−0.93 to −0.64)
Eggs*	—	—
Cereals	−0.87 (−0.99 to −0.74)	−0.72 (−0.85 to −0.59)
Fats and oils	—	—
Sweets, confectionery, and sweetened beverages	−0.87 (−1.06 to −0.70)	−0.73 (−0.91 to −0.55)
Other	−1.06 (−1.21 to −0.92)	−0.93 (−1.08 to −0.78)
All food groups combined	−0.91 (−1.00 to −0.83)	−0.77 (−0.86 to −0.68)

Source: Green et al. (2013), used under the terms of the Creative Commons Attribution 4.0 licence.

which GHG emissions arise from farm and non-farm-related food sectors.

The latest Intergovernmental Panel on Climate Change (IPCC) report suggests that agriculture, forestry, and related land use accounts for approximately 24% of global GHG emissions (Pachauri

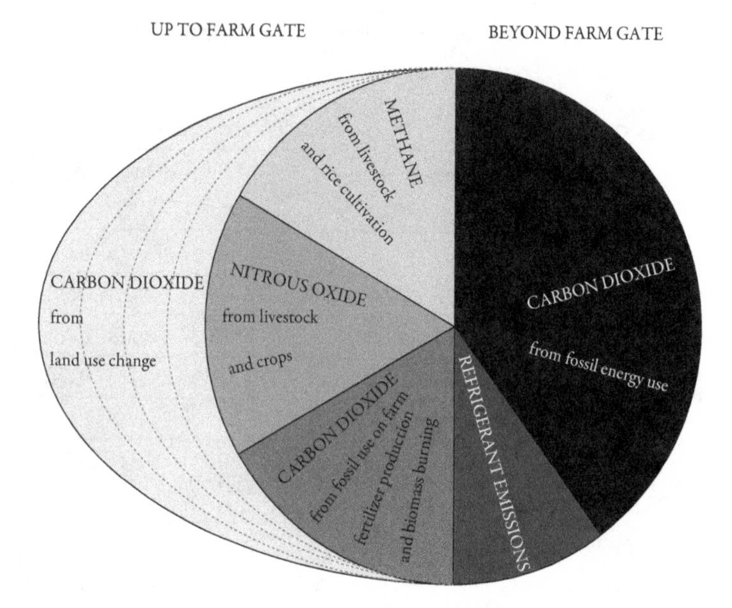

Figure 2.7 Food-chain impacts and the distribution of the different gases. Source: Garnett (2011), used under the terms of the Creative Commons Attribution 4.0 licence.

et al., 2014). While all food systems produce GHG emissions, the industrial food system is hugely polluting and a much larger contributor to climate change than small-scale farming combined with localized distribution networks. The industrial food system is driving deforestation and biodiversity loss, land degradation, water overuse, and pollution (Garnett, 2016). By 2014, more than 70% of freshwater use in most regions of the world was dedicated to agriculture (Khokhar, 2017). Meanwhile, land use change for agricultural production is responsible for 80% of deforestation worldwide (Kissinger, Herold, and De Sy, 2012).

The livestock sector in particular is huge. It is the largest user of agricultural land, occupying 70% of all agricultural land and 30% of the Earth's land surface (FAO, 2006a, Herrero et al., 2016). As well as the land area used for grazing, livestock production requires the cultivation of animal feed crops; over a third of the world's existing arable land is dedicated to feed crop production (Herrero et al., 2016). The livestock sector is also a major contributor to GHG emissions, according to some estimates accounting for 18% of global emissions (FAO, 2006a). Livestock farming produces nitrous oxide from manure and feed crop production, and methane from the digestive processes of ruminant animals such as cows and sheep. CO_2 emissions are due largely to land clearing for further livestock production. Such land clearing has resulted in wide-scale deforestation in Latin America and the Caribbean, and the conversion of land for farming generally is the leading cause of tropical deforestation (Balmford, Green, and Phalan, 2015). Figure 2.8 maps the GHG emissions from livestock in different countries over the decade from 1995 to 2005.

Given that the environmental impact of the livestock sector is so damaging, it is very concerning that the global food economy is increasingly driven by a shift in diet and food consumption patterns toward livestock products. Not only is this bad for the environment, we know that a diet high in animal products is also not good for health (Garnett, 2016).

Global per capita consumption of livestock products has more than doubled in the past 40 years. In the developing countries of Asia, consumption of meat has been growing by more than 4%

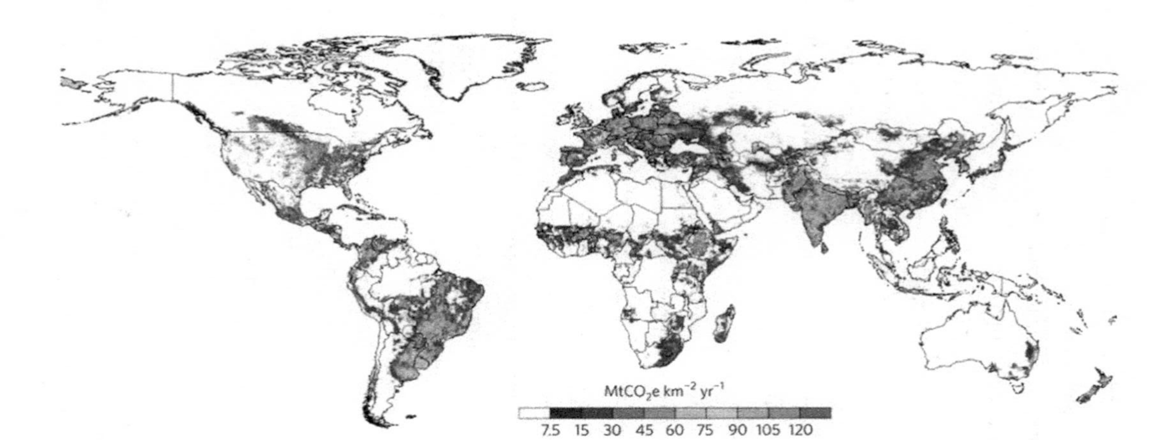

Figure 2.8 Greenhouse gas emissions from global livestock for 1995–2005. Source: Herreo et al. (2016), used under the terms of the Creative Commons Attribution 4.0 licence.

per annum and of milk and dairy products from 2 to 3% per annum. The expansion of the human population from about 7.6 billion in 2017 to an estimated 9 billion by 2050, combined with rising incomes and rapidly rising per capita demand for animal protein—especially in emerging economies—means that total agricultural demand is likely to double between 2000 and 2050 (FAO, 2006b). Beef and milk production have already more than doubled over the past 40 years. Intensification of production has played a pivotal role in this. As Herrero and colleagues (2016) note, "in the USA, 60% more milk is produced now than in the 1940s with about 80% fewer cows." These trends are driving increases in GHG emissions, deforestation, biodiversity loss, land degradation, water overuse, and pollution (Garnett, 2016; Herrero et al., 2016).

Eating Oil

As discussed earlier, the price of oil influences food prices, with high energy prices inflating food prices because of the increased cost of inputs such as fertilizers and pesticides, higher transport costs, and the conversion of food crops to biofuel crops as demand for the latter rises. What I want to emphasize in this scenario is how heavily dependent the industrial food system is on fossil fuels (Pfeiffer, 2009; Sage, 2013). As Figure 2.9 illustrates, fossil fuels, but in particular petroleum, are used at all stages in the food system. This means that a food system predicated on fossil fuel use is constantly contributing to GHG emissions and environmental degradation.

Petroleum is required to produce fertilizers and pesticides and to run the machinery on farms. Once foodstuffs leave the

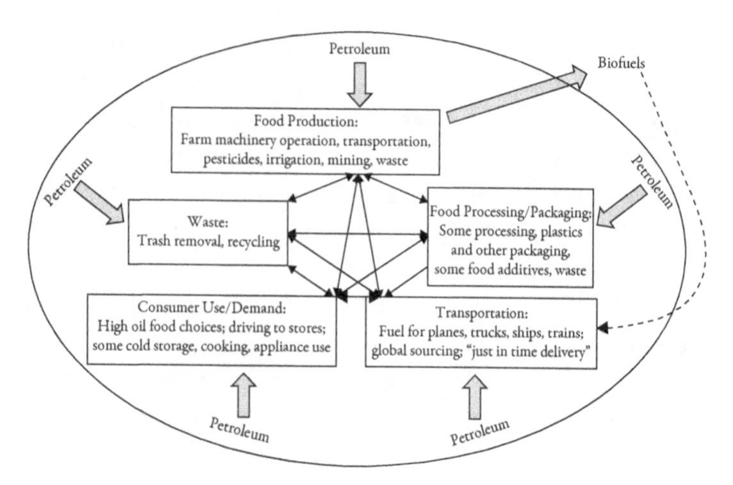

Figure 2.9 Petroleum use in the industrial food system before peak oil. Source: Neff et al. (2011), used under the terms of the Creative Commons Attribution 4.0 licence.

farm, the bulk of food-related emissions come from fossil fuel use for transport, food manufacturing, retail, and across the food service sector. We have seen that within the industrial food system, there is a predominance of energy-dense foods that are highly processed. This usually goes along with food being packaged. Packaged foods have a high environmental impact not only because of production processes but also because of the energy required to make the packaging that enables storage and extends shelf life (McMichael, Butler, and Dixon, 2015; Pretty et al., 2005). Globalized food supply chains also require huge amounts of fuel for planes, ships, and heavy-goods vehicles, while households use their petroleum-filled cars to do weekly food shopping (Pfeiffer, 2009) (Box 2.4).

Box 2.4

"The consumption of fossil fuels, nuclear power, and renewable energy by the US food system was on par, in 2002, with the entire national energy budget for India and exceeded the combined energy budgets of all African nations" (Canning et al., 2017).

Death by Chocolate: Impact of the Industrial Food System on Human Health

The fact that we now live in a world in which almost two billion people consume too many calories but too few micronutrients (International Food Policy Research Institute, 2015) can largely be attributed to the reach and dominance of the industrial food system. There is not a shortage of food among these populations; its nutritional quality is compromised. In industrialized countries, the transition to highly processed starchy foods and diets high in salt, fats, and sweeteners is widespread and continues to gather momentum. In the United States, where more than 75% of all packaged and processed food has some form of added sugar (Anand et al., 2015), it has contributed to what is widely recognized as "an obesity epidemic" (Pinard et al., 2013). A similar nutrition transition is showing marked acceleration in low- and middle-income countries (Popkin, 2017).

Diets containing excessive amounts of highly processed foods are associated not only with obesity but also with increased risks of other—often related—noncommunicable diseases (NCDs) including heart disease, stroke, atherosclerosis, insulin resistance, diabetes, chronic kidney disease, osteoporosis, dental decay, gall bladder disease, and some cancers (WHO, 2003). At the same time as excess consumption is preventing healthy flourishing for two billion people, we saw in Chapter 1 that almost one billion people go hungry and suffer from undernutrition. If the world's primary response to malnutrition continues to be a focus on increasing overall production through the operation of the industrial food system, not only will we fail to halt the rising tide of NCDs, but climate change will ensure the persistence of undernourishment and, indeed, will likely contribute to its rapid increase.

Summary: The Industrial Food System, Climate Change, and Health Inequities

In summary, greater economic globalization and marketing and the increasing power and influence of transnational agriculture, food, and beverage corporations have profoundly altered the functioning of food systems, displacing mainly local systems with a worldwide industrial food system. These shifts have changed the relative balance of foods that are available for consumption toward those that are high in refined sugars, those with refined fats, and meats. Combined with changes in price and heavy marketing, this has resulted in communities' greater willingness to consume highly processed foods (Baker and Friel, 2016; Buse, Tanaka, and Hawkes, 2017; Clapp et al., 2015; Schram et al., 2015). These foods are unhealthy. They, and the system that produces them, are also major contributors to climate change (Foley et al., 2011). The extremes of inequity in our world, and in the food system, are vividly illustrated by the divergence in nutrition outcomes. Far too many people continue to die as a result of undernutrition or to live with illness because of hunger. At the same time, huge numbers of people die or are disabled because they eat too much or because the food they eat is unhealthy.

The Concrete Road to Freedom: Urbanization and the Consumptagenic System

Cities are home to more than half of the world's population, and they continue to expand rapidly. Since the Industrial Revolution, cities have represented places of hope and opportunity for impoverished rural dwellers or those wishing to expand their horizons. For many, they offer an alluring array of work and social opportunities or, in Sen's terminology, places affording the "freedom to lead a life one has reason to value" (Sen, 1999). At the same time, however, cities are engine rooms of economic growth and thus of consumption—powering the consumptagenic system via the demand they create as well as what they produce, and driving climate change and health inequities at a furious rate. Those who have been inclined to think of cities as meccas of choice and freedom have perhaps been blind to the harm done to the planet and human health through processes of industrialization, land use changes, and movement from rural to urban areas in pursuit of a better quality of life.

How consumptagenic urban systems have evolved replicates in many respects the development and globalization of the industrial food system. Beginning mainly in the United States and Europe in the nineteenth century, industrialization, technological innovation, and the attendant increased mining of natural resources—initially close to home but subsequently largely in Africa, Asia, and Latin America—plus international trade brought prosperity to many developed countries and some urban areas in the global South (Lankao, 2007). Along with this trend came the development of urban centers as economic powerhouses and as magnets for people from rural areas (van der Woude, Hayami, and De Vries, 1995).

After a period of intensive state-sponsored investment in agricultural research and development during the 1960s and 1970s, the 1980s saw steep global investment declines in agriculture and rural infrastructure and amenities (United Nations Department of Economic and Social Affairs [UNDESA], 2011). Growing rural poverty and the attractions of employment opportunities and more developed infrastructure and service delivery spurred out-migration to urban centers (McGranahan and Satterthwaite, 2014; WHO, 2016c). Mechanization and intensification of agricultural production, combined with the concentration of land among fewer landholders, is also thought to have played a role in pushing people from the countryside into cities (Clark and York, 2005; Marcotullio et al., 2014).

These urban transformations took place over the course of the nineteenth and twentieth centuries. The current form of urbanization, however, is much quicker and is occurring on a scale not seen before. At the start of the twentieth century, just 1 in 10 people lived in urban areas (Pataki et al., 2011; WHO, 2016c; Zhang et al., 2014), or around 200 million people in total (Steffen et al., 2007). By 1950 this had increased to 3 in 10 and by 2000 to 5 in 10—a total of three billion people (Steffen et al., 2007). By 2030,

6 in 10 people will be urban dwellers, the numbers rising to 7 in 10 by 2050 (International Food Policy Research Institute, 2017; WHO, 2016c). According to UN estimates, more than 90% of future urban population growth will be in low- and middle-income countries (International Food Policy Research Institute, 2017; WHO, 2016c) (Figure 2.10). It is expected that between 2010 and 2050, urban dwellers in Africa will increase by 944 million and in Asia by 1.449 billion (McGranahan and Satterthwaite, 2014).

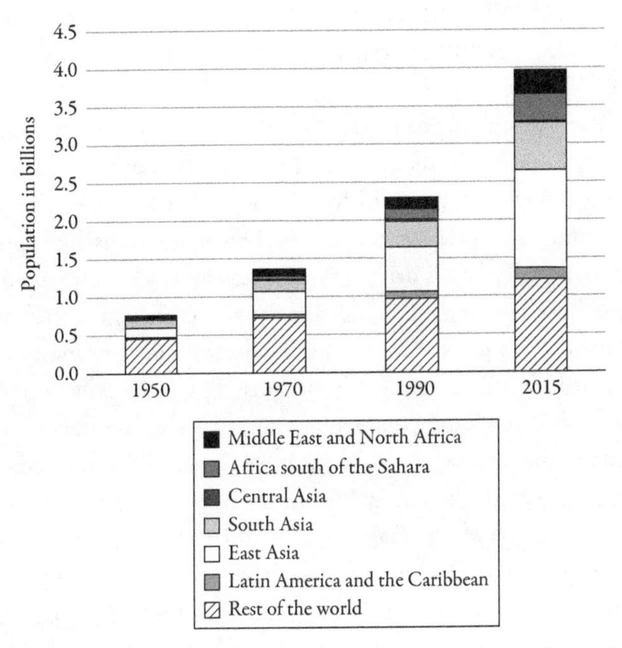

Figure 2.10 Growth of urban population in major developing regions. Source: International Food Policy Research Institute (2017), used under the terms of the Creative Commons Attribution 4.0 licence.

Cities Are Killing the Planet and People, But How They Drive Climate Change Varies between Countries and within Cities

As cities grow, the concentration of population and concomitant changes in social, political, and economic activities accelerate carbon releases through land use change and increased use and consumption of energy and materials (Marcotullio et al., 2014). According to some estimates, cities are responsible for more than 70% (WHO, 2016c)—or a remarkable three-quarters (Seto et al., 2014)—of global GHG emissions. Most of these arise from energy use (WHO, 2016c). On the basis of current trends, Creutzig and colleagues (2015) estimate that by 2050 global urban energy use will triple.

But the amount of emissions produced by particular cities is highly variable. Considering the role of cities as drivers of climate change is also complicated by the methods used to account for emissions. As noted earlier in this chapter when discussing how international trade is driving GHG emissions, while the standard procedure is to attribute emissions to the place where they were generated, scholars who emphasize the role of consumption in the creation of emissions call for consumption-based accounting methods—attributing emissions associated with production of a good to the place where it is consumed rather than produced.

Unlike before, twenty-first-century cities are highly connected and reliant on the global economy. Most commodities are not made and sold locally, and even services are now commonly supplied from a distance. These goods and services are part of a global supply chain, coming from other cities—frequently in the global South—where economies of scale and comparative advantage ensure the greatest capital returns for the transnational corporations engaged in production, manufacture, and supply. This means that the GHG emissions traceable to urbanization come not only from the energy use associated with transportation, buildings, and industry within particular cities but also from the global movement of goods and services.

Cities in affluent countries drive GHG emissions because they operate as hubs of demand-fueled consumption. This consumption is itself fueled by the ability of manufacturers to drive down production costs by locating—and endlessly relocating—production to wherever in the world is cheapest in terms of labor and other costs. According to standard methods for attributing emissions, the effect of this "off-shoring" of production processes is to lower the GHG emissions attributed to the cities where the goods and services are consumed. But even though this is the case, cities in affluent countries still produce much higher emissions than cities in poorer countries (UN-Habitat, 2008). Analysis by Hoornweg and colleagues suggests that average per capita GHG emissions for cities vary from more than 15 tons of carbon dioxide equivalent in wealthy cities, including Sydney, Calgary, Stuttgart, and a number of major US cities, to less than half a ton in many cities in relatively poor countries including Bangladesh, Nepal, and India (Hoornweg, Sugar, and Trejos Gómez, 2011) (Figure 2.11). Other factors influencing the relative contributions of different cities to GHG emissions are density and transport infrastructure. Dense cities have lower emissions than cities that are more spread out, and cities with extensive and affordable public transport—or in which most residents do not own cars—produce far less emissions than those that are highly car dependent. Car dependence and the sprawling

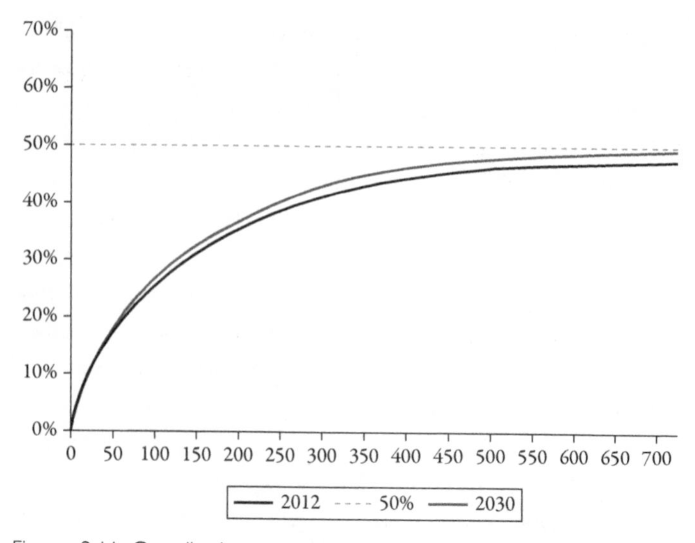

Figure 2.11 Contribution to global carbon emissions by cities above 0.5 million. Source: LSE Cities (2014), used under the terms of the Creative Commons Attribution 4.0 licence.

nature of cities in the United States combine to ensure their GHG emissions are "exceptionally high" (UN-Habitat, 2008).

Aside from the variability in emissions between different cities, there is also a high level of variability in the level of emissions produced by different areas within cities. Local-level affluence correlates strongly with higher emissions, although little systematic data have been collected to allow detailed mapping of this (UN-Habitat, 2008).

Urban Settlements as Determinants of Health Inequities

Urbanization drives economic growth and has led to greater prosperity, particularly within urban areas themselves, which are generally more prosperous than rural areas (Chen et al., 2014; Floater et al., 2014; WHO, 2016c). Many cities have seen marked improvements in living conditions over time, with improvements in housing and sanitation, average household income, levels of education, and opportunities for women to participate in the labor force (Turok and McGranahan, 2013; WHO, 2016c). But the economic and social development effects of urbanization are variable. In a number of cities, the urban transition has created unmet demand for housing, transport, work, and social and health services, and a growing share of world poverty is in urban areas (International Food Policy Research Institute, 2017; McGranahan and Satterthwaite, 2014; Tacoli, McGranahan, and Satterthwaite, 2015; WHO, 2016c).

All of this matters for health and health inequities. Within cities in many wealthy as well as poorer countries, health inequities are growing. In Chapter 1, I mentioned the 15-year life expectancy gap between privileged and underprivileged communities in Glasgow, Scotland. In London, England, this life expectancy gap is 17 years, and in Baltimore, Maryland, in the United States, there is a 20-year life expectancy gap between the city's most and least privileged residents (WHO, 2016c). Internationally, in all four regions shown in Figure 2.12, children in the poorest 20% of urban households are at least twice as likely to die as children in the richest 20% of urban households (WHO, 2016c).

The most extreme forms of urban poverty are concentrated in informal urban settlements. Reflecting their precarious legal status and virtual absence of services such as potable water and waste removal, these settlements are often referred to as "slums." Recent estimates provided by the World Health Organization show that the proportion of urban populations living in slums in

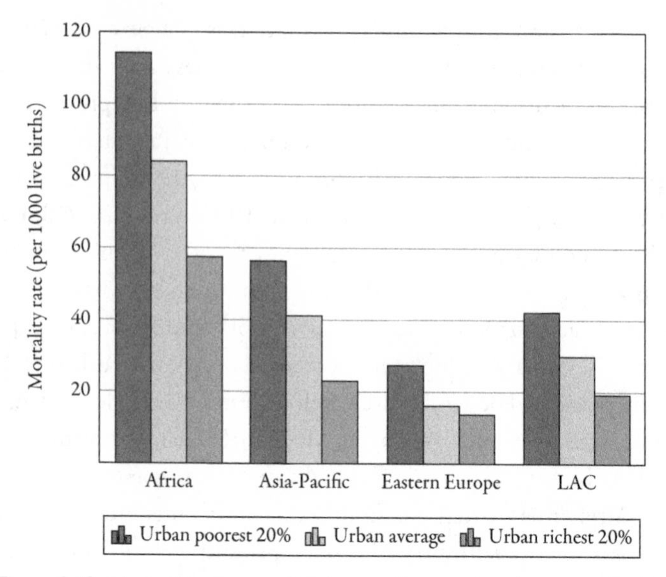

Figure 2.12 Under-five (years of age) mortality rate in urban areas, by UN region and urban wealth quintile, 2005–2013. LAC, Latin America and the Caribbean. Source: WHO (2016a), with permission to display from World Health Organization.

the developing world decreased from 46.2% in 1990 to 29.7% in 2014, but these figures are not uniform across regions. In many low-income countries and throughout sub-Saharan Africa, a majority of the urban population lives in slums (Kjellstrom, 2009a; WHO, 2016c). In absolute terms, the number of slum dwellers has increased, with nearly a billion people living in slums in 2016, compared to 689 million in 1990 (WHO, 2016c). The World Health Organization predicts that this figure will reach two billion by 2050 (WHO, 2016c) (Figure 2.13).

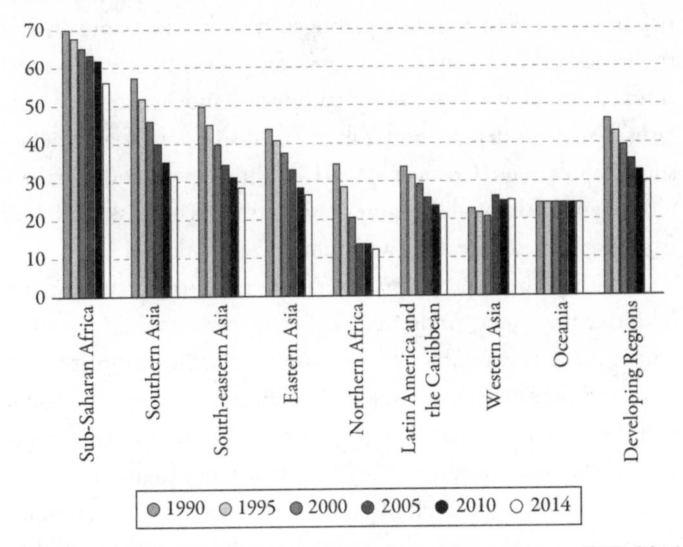

Figure 2.13 Percentage of urban populations living in slums, 1990–2014. Source: UN-Habitat (2016), used under the terms of the Creative Commons Attribution 4.0 licence.

Poor urban living conditions, particularly for people living in informal settlements, are breeding grounds for communicable disease (Eckert and Kohler, 2014). The makeshift housing, overcrowding, and inadequate water supply, sanitation, drainage, and garbage collection that typically characterize urban slums expose people who live there to recurrent diarrhea and diseases such as typhoid, hookworm, and cholera and are major risk factors for mortality (Mberu et al., 2016). Given the nature of the living environment, slum dwellers are at increased risk for malaria and dengue, both climate-sensitive health conditions. They are also at increased risk of exposure to zoonotic pathogens that live in mammals and to viruses spread by mosquitos. As exemplified by urban centers throughout Brazil, rat-borne pathogens and mosquito-borne viruses such as dengue and Zika virus disproportionately affect slum populations (Costa et al., 2017). The physical and social environment of slums also exposes residents to health risks of injury from fire, extreme weather, and crime (Ezeh et al., 2017).

Children are especially vulnerable in slums because of low breastfeeding rates, undernutrition, and poor sanitation, all of which make them vulnerable to chronic diarrhea, stunting, and impaired cognitive development. Evidence from a pooled sample of 1.38 million children across 73 low- and middle-income countries shows that children living in slums suffer higher mortality rates and stunting than children who live in better-off areas of the same urban settlements but lower mortality rates than children living in rural areas (Fink, Gunther, and Hill, 2014). In relation to acute illnesses such as diarrhea, however, children living in slums confront higher risks than do children living in better serviced urban areas and children who live in rural areas. A large fraction of the observed urban health inequities could be explained by inequities in maternal education, household wealth, and access to health services across residential areas (Fink et al., 2014).

Just as communicable diseases harm and kill residents in many cities in low- and middle-income countries, urbanization, another driver of the global nutrition transition toward processed foods that are high in sugar, salt, and fat (International Food Policy Research Institute, 2017), brings with it an acceleration in rates of NCDs including diabetes, heart disease, and obesity. Urbanization is also associated with respiratory conditions, mental health problems, alcohol and drug abuse, and violence (Ezzati et al., 2005; International Food Policy Research Institute, 2017; Ng et al., 2014; Wang et al., 2016). Some of the most populous regions and countries in the world face an epidemic of NCDs. Low-income countries are affected most, with death rates inversely proportional to a country's gross national income. Angkurawaranon and colleagues (2014) found that in countries across South East Asia, urban living correlated with coronary heart disease, diabetes, and respiratory diseases in children, but with reduced rates of rheumatic heart diseases. Others have found that the rapid rise globally in people who are overweight and obese is concentrated in urban areas (Goryakin and Suhrcke, 2014; International Food Policy Research Institute, 2017).

Cities, Climate Change, and Health Inequities

In many respects, urbanization and climate change combine to create a perfect storm. Once climate change is added into the mix with the factors just discussed, the scale of the health impact of urban living is enormous and will amplify existing inequities within countries and internationally (Friel et al., 2011; Rydin et al., 2012). As we saw in Chapter 1, while everyone will be touched by climate change, some will be affected more than others. One major divide is between the global North and South, with cities in low-income countries in the global South likely to suffer the worst effects (Reckien et al., 2017). Within cities in high- and middle- as well as low-income countries, the health-related impacts of climate change will be greatest for members of low socioeconomic groups (Reckien et al., 2017). The impact of climate change on health-related inequities between and within cities will be both direct and indirect, as visualized in Figure 2.14 (Friel et al., 2011).

Reckien and colleagues suggest that the risks of climate change in urban settlements differ in accordance with (1) physical exposure, determined by the location of the urban settlement and of subpopulations within it; (2) social, economic, and demographic characteristics; (3) institutional and governance characteristics; and (4) urban development processes that "construct risk." These interrelated factors can be seen at play when analyzing the effects of heat, sea level rise, and inland flood risks on urban health inequities (Reckien et al., 2017).

Heat, Living and Working Conditions, and Urban Health Inequities

Higher temperatures are a major health issue for urban dwellers, with cities experiencing both the overall increase in global average temperatures caused by climate change and localized increases caused by the urban environment itself. The "urban heat island effect" results from a lack of shade and vegetation in cities, and a density of road and building surfaces that absorb and retain heat (Watkins, Palmer, and Kolokotroni, 2007). While higher temperatures may have health benefits in very cold climates, excessive heat and heatwaves have caused increases in mortality and morbidity in cities in Korea (Son et al., 2012), the United States (Anderson and Bell, 2011; Chow, Brennan, and Brazel, 2012; IPCC, 2007; Luber and McGeehin, 2008), Europe (American Psychological Association, 2010; D'Ippoliti et al., 2010; Robine

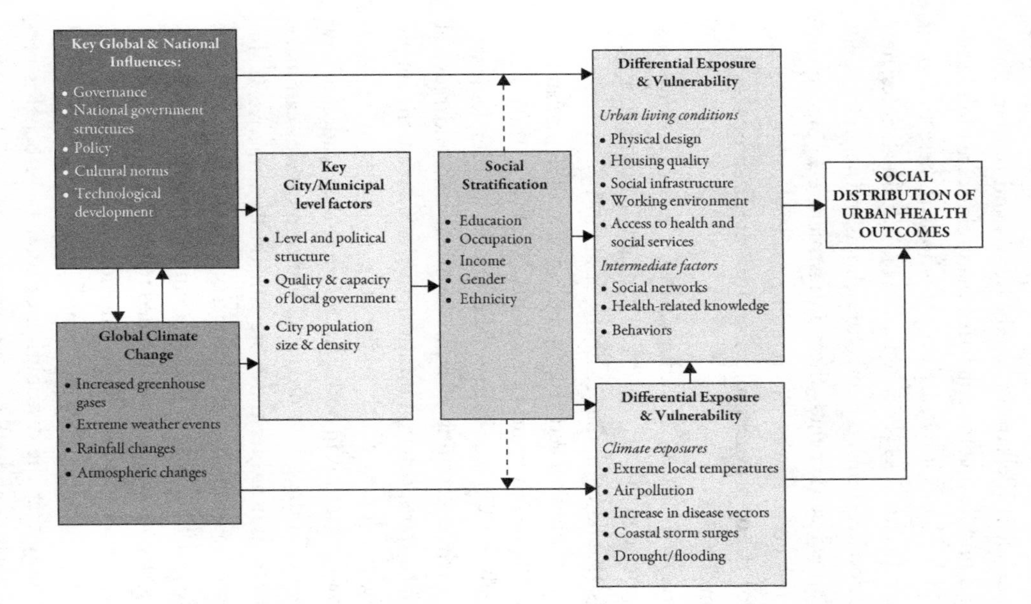

Figure 2.14 Direct and indirect pathways from global climate change via urbanization to health inequities. Source: Friel et al. (2011), used under the terms of the Creative Commons Attribution 4.0 licence.

et al., 2008), and many developing countries (Egondi et al., 2012; Hajat et al., 2005; Kjellstrom, 2009b; Kovats and Akhtar, 2008). In low- and middle-income countries, many cities are built in hot and humid lowlands, exposing people who live there to serious heat-related health risks (Friel et al., 2011). Within all cities, heat disproportionately impacts poorer households and neighborhoods and unplanned settlements where infrastructure is inadequate, there are few green spaces, and building materials often intensify the heat or fail to provide protection from it (Ezeh et al., 2017; Kjellstrom, 2009a).

Urban green spaces, including parks, forests, green roofs, streams, and community gardens, cool their immediate surroundings (Pataki et al., 2011; Zhang et al., 2014). Green space also promotes physical activity, psychological well-being, and the general public health of urban residents (WHO, 2016d). In the United States and China, a large number of cities have implemented strategies such as greening of remnant urban land and reuse of underutilized transportation infrastructure to increase the supply of urban green space. Wolch and colleagues (2014) note, however, a growing paradox in both places: "while the creation of new green space to address environmental justice problems can make neighbourhoods healthier and more esthetically attractive, it also can increase housing costs and property values. Ultimately, this can lead to gentrification and a displacement of the very residents the green space strategies were designed to benefit".

In their review of the Anglo-American literature on urban green space, especially parks, Wolch and colleagues (2014) found more generally that the distribution of such space benefits predominantly white and affluent communities. A study by Uejio and colleagues (2011) of heat exposure and the built environment in the US cities of Phoenix and Philadelphia similarly found that more heat-related distress calls were made from neighborhoods with higher proportions of black, Hispanic, linguistically isolated, and renting residents. The areas of Philadelphia and Phoenix with the most extreme health problems caused by heat stress are those that have experienced economic stagnation and Anglo-American out-migration. Uejio and colleagues argue that discriminatory economic practices prevent people of color from living in cooler, safer, and newer housing in each city.

Climate change is also exacerbating health inequities by making working conditions in urban areas more difficult, especially for the underprivileged. Analysis by Kjellstrom and colleagues has shown that, later this century, many among the four billion people who live in hot areas worldwide will experience reduced work capacity owing to climate change. In some areas, 30–40% of annual daylight hours will become too hot for work to be carried out. The social and economic impacts will be considerable: the potential for global gross domestic product losses is greater than 20% by the end of the century (Kjellstrom et al., 2016). Industries whose workers are frequently exposed to excessive heat in urban areas include construction, services, manufacturing—especially where it involves outdoor job activities or is in confined areas—and transportation. Many factories and workshops lack air conditioning, particularly those in low- and middle-income countries. The indoor heat-exposure levels of people in most of the tropics are already high, and they are increasing (Hyatt, Lemke, and Kjellstrom, 2010). For individuals working in lower occupational grades who already experience worse health outcomes than workers in higher-graded occupations this disadvantage will be exacerbated by temperature extremes, especially among individuals working in the industries just listed (Kjellstrom, 2009b).

Living the Coastal Dream: Risks to Urban Health Equity from Sea Level Rise and Storm Surges

As we saw in Chapter 1, other major threats to the survival and health of people living in cities around the world are sea level rise and storm surges, with many cities, including 15 of the world's 20 largest, located in low-lying coastal zones and population growth concentrated in these areas (Dodman et al., 2013; Neumann et al., 2015; World Bank, 2010a). Exactly how high the sea level will rise is an ongoing debate (Howes et al., 2015). The IPCC expects sea level rise to accelerate and predicts rises between 0.26 and 0.82 meters by 2100 (IPCC, 2014b).

Strauss and colleagues, at Climate Central, consider the possible sea level rises *beyond* 2100 on the basis of climate warming likely to occur this century. They estimate that climate warming of 4°C above pre-industrial levels (which they describe as likely on a "business as usual" approach, consistent with IPCC assessments) will lead to a median sea level rise of 8.9 meters. On the basis of 2010 population figures, this would flood areas that are home to around 627 million people. By comparison, a 2°C temperature rise is likely to produce a median sea level rise of 4.7 meters, inundating land that was home in 2010 to 280 million people (Strauss, Kulp, and Levermann, 2015).

The regional impact of sea level rise is uneven, with 83% of people likely to be affected living in less developed countries (Neumann et al., 2015) and 74% of these people concentrated in Asia (Strauss et al., 2015). The potential scale of devastation is enormous—Mumbai saw massive floods in 2005, as did Karachi in 2007. More recent examples of coastal flood disasters include Hurricane Sandy in 2012 in New York City, and Super Typhoon Haiyan in 2013 in the Philippines. Shanghai has been declared the most vulnerable major city in the world to serious flooding. Other megacities that are highly vulnerable include Hong Kong, Calcutta, Mumbai, Dhaka, Jakarta, and Hanoi (Strauss et al., 2015).

Sea level rise, combined with increased storm surge, puts the lives of people living in low-lying cities at risk from flooding and resulting damage to infrastructure, including roads, housing, and water and sanitation systems. Poorer urban households are usually at higher risk because their homes are structurally less robust, because they tend to live closer to sea level, and because the infrastructure in poorer cities is less able to withstand damage (Costello et al., 2009; Dodman et al., 2013). Women may be more affected because they are more likely to work in home-based businesses. Poorer urban households often lack the economic resources to evacuate in the face of climate-related disasters or to rebuild damaged homes and infrastructure (Reckien et al., 2017). The flooding of New Orleans in 2005 and its devastating impact on elderly nursing home patients and poor people who could not evacuate because of lack of transport is a striking example of how climate change can affect socially disadvantaged urban communities even in rich countries (Sharkey, 2007).

City Living Inland: Flooding and Health Impacts

It is not only coastal city dwellers who are at risk from climate change and its associated health effects. Inland cities are also vulnerable to flooding as climate change increases the likelihood of torrential downpours. Flood risk in cities is heightened by impermeable surfaces that create rapid run-off; inadequate drainage systems; and by construction on marshlands and floodplains (Reckien et al., 2017). An estimated half a billion urban residents in Asia live in substandard housing or informal settlements, often adjacent to rivers and canals. In Africa, around 411 million people currently live in informal settlements. Given the proximity of many of these to waterways, the urban poor risk losing their homes to flooding, often leading to their permanent displacement (Reckien et al., 2017). This also implies loss of livelihoods and community support networks, with major implications for mental health, as noted in Chapter 1 (McMichael et al., 2012; Reckien et al., 2017).

Other health risks arise from unsanitary conditions. As discussed earlier, people living in informal settlements already lack water and sanitation of a standard that is healthy and convenient (WHO, 2016c). They do not have sewers or covered storm drains, or if they do these only serve a small percentage of the population. They do not have household waste collection services or street cleaning. When infrastructure is inadequate, levels of poor sanitation and drainage and impure drinking water increase during flooding, leading to the transmission of infectious diseases. Outbreaks of cholera, dysentery and diarrheal diseases, acute respiratory infections, dengue, and malaria peak in the dense low-income neighborhoods of cities following intense and excessive rainfall (Bezirtzoglou, Dekas, and Charvalos, 2011; Reckien et al., 2017; Revi, 2008).

Summary: The Relationship between Urbanization, Climate Change, and Health Inequities

In this part of the chapter I posed the provocative idea of urbanization as the concrete road to freedom. What my analysis of urbanization presented, sadly, is the reality that not all social groups enjoy that freedom. Economically and socially disadvantaged households in cities throughout the world already experience poor health outcomes compared with more affluent households. Once climate change is added to the mix, these inequities widen, and people whose lives are already fraught face increased and often acute physical and mental health risks. The environmental challenges associated with urbanization are now considered a global development priority. In 2016, 167 countries at the UN Habitat III summit adopted the "New Urban Agenda," dedicated to promoting sustainable development (International Food Policy Research Institute, 2017). The Sustainable Development Goals also include the aspiration to "make cities inclusive, safe, sustainable and resilient" (SDG 11). Achieving these aims requires addressing the consumptagenic nature of urbanization and integrating measures to limit global warming with an immediate and concentrated focus on reducing existing inequities, including health inequities, in urban environments.

Concluding Remarks

The analysis in this chapter, plus the discussion in Chapter 1, highlights that the relationships between climate change and health inequity, and the underlying social determinants that fuel them, are not straightforward; they are messy and complex. I pick up on these ideas of complexity in Chapter 3, arguing that how we understand and characterize these phenomena depends on our standpoint and also evolves as knowledge develops and available evidence accumulates. Given these complexities, responding to climate change and health inequity requires the development of policies that transcend doctrinal boundaries and draw on insights from a range of disciplines, including those in the sciences, social sciences, and humanities.

3 | CHALLENGES AND FUTURE PROSPECTS

Introduction

Chapter 2 described the evolution of the consumptagenic system and its addiction to conspicuous overconsumption. It described how the industrial food system and urbanization drive forms of consumption that are unhealthy and also contribute to climate change, creating feedback loops that further deepen health inequities. This final chapter posits that for the research, advocacy, and policy worlds to move forward in a way that addresses environmental degradation, social justice, and health equity, they must break down disciplinary, sectoral, and policy silos and think and work in systems. This is a challenge to the current orthodoxy in public health research and policy.

This chapter begins by identifying two key gaps in the evidence base. The first involves systems research. I explain what systems thinking is and show why resorting to it is crucial as a means of identifying the problems with which the book is concerned and for crafting solutions. The second key gap is focusing on the question of how to mobilize action and achieve targeted and effective outcomes. I urge a move away from research based primarily on reductive pathologizing and toward research that is attentive to, and actively engaged with, policymaking processes.

The next section of the chapter steps back to ask what kind of world we are seeking to achieve. With an understanding of the vision we are pursuing, the chapter examines how progressive policy systems might be created and deployed to reign in consumption. It lays out some options for intersectoral action designed to achieve greater equality, environmental sustainability, and health equity. Understanding that there are common determinants of climate change and health inequities provides an opportunity to "kill two birds with one stone" through the appropriate design and implementation of policies and actions across a range of sectors.

In this section, I argue that the blind pursuit of economic growth must be replaced by an enlightened commitment to greater material equality. Existing policy responses ignore or sideline widening material inequalities within states and internationally, yet a growing body of research demonstrates that more equal societies are not only better for the individuals within them across a wide range of indicators, including health and well-being, they are also more environmentally sustainable. Targeted actions to redress material inequalities produce systemic effects that are essential preconditions for effective action on climate change, as well as for the achievement of health equity. This part also reviews some of the existing policy responses to climate change and health inequity first discussed in the book's Introduction, including the UN Agenda for Sustainable Development, the approach advocated by the Commission on Social Determinants of Health, and the 2015 Paris Climate Agreement.

The chapter concludes with the acknowledgment that resistance to change will likely be encountered, but also with the hope that a plurality of disciplinary approaches, systems science, and a greater understanding of policy processes will bring the book's vision for a healthier and more sustainable world into clearer view.

A Fit-for-Purpose Evidence Base

Each of us concerned about human health, environmental sustainability, and social justice are model mongers and chaotic innovators—mobilizing and working in different ways and in different venues toward a vision of healthy, fair, and just systems (Friel, 2018). Researchers provide important evidence, which may be ignored or buried by some. But it is not ignored by others and it is not ignored all the time. Evidence is power. As a health-concerned research community we must ask ourselves if we are making the most of that power. Do we use the complement of theoretical, methodological, and empirical research tools that most effectively communicate the importance of the climate change and health equity issues our world confronts; that demonstrate the many opportunities, and need, for action across a suite of policy domains and that show pathways forward for coherent policy development and implementation? I think we do not. A research agenda concerned about climate change and health equity would benefit from a plurality of approaches, in particular more attention to systems-oriented analytical approaches, and also a much greater focus on understanding the policy, including advocacy, processes that enable or impede political and policy attention to these issues.

Thinking in Systems

What Chapter 2's exposé of the consumptagenic system and consumptagenic societies demonstrates, alongside the analysis in Chapter 1, is that there is a plethora of interconnected factors—environmental, social, economic, and health related—that operate in multidirectional and nonlinear ways to influence climate change, social conditions, and health inequities. To fully understand and, ultimately, act on issues that will improve health equity and reduce climate change, we must think in systems.

As noted in Chapter 1, developments in epidemiology have expanded the theoretical and analytical framework used by the discipline, and there is now greater recognition of the different levels affecting health outcomes, including macro-, meso-, and micro-levels; types of causal factors; different disease trajectories and outcomes; and notions of temporality (Krieger, 2001; Susser and Susser, 1996). We now know more than ever before about how climate change interacts with the social determinants of health inequity, exacerbating these and other social inequities (McMichael, 2017). This emergence of eco-epidemiology is to be applauded. There are, however, still limitations with this methodological approach because it is not designed to capture complex interactions between the drivers of climate change and health inequities. These interactions can happen in unexpected ways and can produce outcomes that are counterintuitive or hard to predict.

Systems science developed as a way of organizing and analyzing complex information in a manner attentive both to the overall operation of systems and to the dynamic interaction of variables within them. With roots in mathematical theories of nonlinear dynamics, it has been adapted and applied in diverse disciplinary settings (Atkinson et al., 2015; Forrester, 1961, 1969; Richardson, 2011; Sterman, 2000, 2006). Since the mid-1950s, systems science has been widely used to inform policy and practice, including in ecology, strategic studies, international relations, economics, engineering, and business (Richardson, 2011).

A *system* refers to a collection of elements—for example, actors, subsystems, sectors—and the interconnections between them that give rise to dynamic behavior (Meadows, 2008; Proust et al., 2012; Sterman, 2000). Like more familiar socioecological approaches, a systems-based perspective is attentive to structural causes of health inequities. It "go[es] beyond the biology and behaviour of the individual human being" to consider the role of social, economic, political, and natural environments in influencing health outcomes (Midgley, 2006, 466). The value of calling an observed situation a "system," however, is to emphasize that it is not possible to understand the phenomenon by studying its parts or elements in isolation; attention to the dynamics between the parts is fundamental (Proust et al., 2012).

There is now considerable evidence about the character of human societies as complex adaptive systems. These social systems exist within other interdependent systems, such as the Earth system. As described later in this chapter, complex adaptive systems are dynamic, self-organizing, and display messiness, novelty, and learning. From a systems perspective, climate change and social and health inequities are considered to be the emergent properties of complex adaptive systems (Meadows, 2008). The notion of emergence refers to "the arising of novel and coherent structures, patterns and properties during the process of self-organisation in complex systems" (Corning, 2002, 49, in Goldstein, 1999). Properties that emerge from systems involving humans do not do so passively. Rather, human decision-making influences how the system is structured and the outcomes it produces, such as climate change or health inequities. Actors are therefore central in such systems, whether these actors are policymakers, business people, public health professionals, researchers, or members of the public. They engage in multiple actions, practices, and patterns of behavior, linked in complex relationships with each other and with their social and material world. Given the complexity of relationships and the number of variables within systems, their impacts and outcomes are difficult to predict.

The adaptive aspect of a complex adaptive system reflects the changing nature of the system and again demonstrates the agency of actors within it, as they have the ability to learn from experience and change the system (Sterman, 2006). Understanding how systems change is critical, and is at the core of complex systems analysis (Richardson, 2011). By studying the endogenous behavior of a system, researchers can gain insight into its core driving processes and identify effective points of intervention based on the levels of agency of the actors within it. Such knowledge could provide real opportunities for effective climate-change adaptation and mitigation in ways that improve health inequities.

Another defining feature of systems understanding is the recognition that multiple interactions between system components are governed by feedback (Holland, 2006; Levin et al., 2012). Feedback is a circular chain of causation that may act to perpetuate or reinforce the state of the system, which in turn leads to emergent properties such as climate change and health inequities. A change in one variable can trigger a string of causal connections that result in either amplifying or opposing the change in another variable. Simply put, change in one part of the system can cause changes in other parts. An example of feedback might be the availability of healthy food promoting a healthier diet, which in turn creates greater demand for healthy foods. Partly because of the inadequacy of traditional quantitative methods in accounting for feedback loops, feedback has been ignored in most public health research. Yet its presence can give rise to nonlinear effects, large effects from small changes in initial conditions, and time delays in the influence of one variable over another, all of which can result in unintended consequences (Meadows, 2008; Sterman, 2000).

It is this reality that has increasingly caused public health professionals and researchers to call for a more systems-oriented analysis to sit alongside traditional approaches to public health and contribute to understanding the problems and ways forward (Bar-Yam, 2006; Carey et al., 2015; Diez Roux, 2011; Joffe and Mindell, 2006; Leischow et al., 2008; Mabry et al., 2010; Milstein et al., 2007; Newell et al., 2007; Rickles, Hawe, and Shiell, 2007; Rutter et al., 2017; Sanson-Fisher et al., 2014; Swinburn et al., 2011). One early initiative pioneering a systems approach was the UK's Foresight Study, which worked across sectoral boundaries to create a systems diagram of the variables that contribute to obesity (Government Office for Science, 2007). A subsequent study by Allender and colleagues (2015) (Box 3.1) demonstrated the effectiveness of using systems methods to identify the determinants of

Box 3.1

Systems Approach to Childhood Obesity Prevention

The study by Allender et al. (2015) demonstrates the power of systems approaches in relation to childhood obesity prevention. The coming together of epidemiologists, systems modelers, and public health nutrition researchers, along with policymakers and community groups in a rural setting in Australia, enabled an analysis of the multiple layers of determinants of childhood obesity that was very contextually relevant and rich. Participants in community modeling workshops represented a variety of interests and priorities, including local planning, sports club participation, providing and/or selling food to children, water and sanitation, and business viability. After being presented with the epidemiology of childhood obesity in their communities, the participants started to identify direct and indirect connections between possible causal variables, culminating in the production of a visual representation of the interconnections between variables. Participants appreciated having a visual representation of the obesity problem because it highlighted the complexity but was easy to follow, linkages were clear, and it acknowledged and displayed the whole story. Participants very quickly moved to identifying solutions. This study demonstrates a robust, systematic process to engage community members in developing an understanding of systems that affect childhood obesity. The approach explicitly engages the community in all steps of the process and provides the basis for subsequent interventions to be owned and driven by that community. A strength was that the approach engaged people at various levels in the community, from leaders to those involved in service delivery. This model has the potential to enhance community readiness for change because the political will, leadership, and workforce are all on the same page and ready to go.

childhood obesity within local communities, and a similar analysis was applied by a team working with this book's author to identify policy entry points for reducing inequities in healthy eating (Friel et al., 2017).

This is by no means mainstream research, however, and certainly not in relation to climate change and health inequities. Redressing health inequities and climate change requires public health researchers to break free from a reductionist paradigm and to think and work in systems. While systems science is not the be-all-and-end-all, it does offer a methodological toolkit very relevant for climate and health policy challenges. By including systems science in their array of acceptable methods, researchers within the field of public health are able to draw on the wealth of knowledge about how complex adaptive systems function and to organize and analyze complex information with an emphasis on the whole picture as well as the interactions between variables.

Moving Beyond the Problem: A Focus on Politics, Policies, and Processes

Public health, and in particular epidemiology, has done an outstanding job of shining a spotlight on the policy problems that confront us in the form of climate change and health inequities. As seen in Chapter 1, there is a wealth of empirical evidence showing how inequities in everyday living conditions in childhood, family life, education, employment, the built environment, and healthcare contribute to inequities in physical and mental health outcomes. There is also increasing epidemiological evidence that people's daily living conditions are affected by structural inequities, which are reproduced through social, cultural, and economic processes. On top of this, we have started to see the demonstration of pathways from climate change to health risks and outcomes, although there is still relatively little epidemiological attention to the relationship in terms of inequities. We need more of that.

What all of this evidence points to is a need for coherent intersectoral policy development and implementation. But getting effective action to improve health and reduce health inequities has been slow and uneven (de Leeuw, 2017; Shankardass et al., 2012). The lack of effective multisectoral policy and action reflects the fact that focusing on the environmental and social determinants of health and on issues of equity challenges established political and policy assumptions and current institutional norms and practices (Baum and Friel, 2017; Baum et al., 2018). Policy processes reflect the ways in which power is distributed in societies, from the initial stages of getting an issue onto the policy agenda through to policy formation, implementation, and evaluation. The involvement of a wide range of actors, including politicians, policymakers, and community and business groups with differing and sometimes conflicting objectives, goes to the heart of the raw politics of power (Raphael, 2015). As Chapter 2 highlighted, the past three decades have seen the ascendancy of market-based policies in many countries, and the reliance of these policies on a strong ideology of individualism has made focusing on broader social determinants more difficult (Baum and Fisher, 2014). This challenging policy environment is compounded by the complexity of the determinants-of-health inequity, making it difficult to allocate responsibility, find coherence between policy goals, and attribute the contribution of particular policies to changes in the distribution of health outcomes.

While evidence is most certainly not the only factor that influences policy decisions, it does matter, even more so in a world in which some political figures disdain it or refute its existence. With the support of a compliant media, politicians and other public figures have considerable capacity to deny or distort reality, but the ability to outrun reality has its limits. Disagreeable facts tend to catch up with us all, and policy for a humane world is best devised on the basis of factual understandings and evidence that are as robust as possible. It might be fairly argued, however, that too much of the currently available public health evidence is at the technical level, focused on "pathologies"—for example, the facts of climate change and of health inequities—rather than on an understanding of the political, policy, and social processes that variously enable or hinder remedial action (Catford, 2009; Friedman and Gostin, 2017; Horton, 2018).

Moreover, many of the conditions and processes that make for effective policy and action are poorly understood theoretically and practically. It is vitally important that, as a research community concerned about climate change and health inequities, we produce evidence that is savvy about policy and political processes and realities (e.g., Box 3.2). Let me shamelessly plug a research program that I am involved with, the Centre for Research Excellence (CRE) in the Social Determinants of Health Equity, funded for 5 years through the Australian National Health and Medical Research Council. The aim of the CRE is "to provide evidence on how political and policy processes could function more effectively in order to operationalise the social determinants to achieve better and more equitable health outcomes." We hope that the work of the CRE is taking research on the social determinants of health equity to a new level, by recognizing the inherently political nature of the uptake, formulation, and implementation of policy. For some papers about the CRE see Baker et al. (2017), Baum and Friel (2017), Baum et al. (2018), Lee et al. (2018), and Schram et al. (2018). Similar research is needed for climate change and health equity.

In essence, I am suggesting that a better balance is needed in research between a focus on "the problem" and a focus on "mobilizing change." Shifting the emphasis will provide multiple pathways forward for coherent policy and action and provide hope that the status quo can change. Such a move away from a reductionist pathologizing requires public health researchers

Box 3.2

Advocacy and the Soda Tax

Analyses of the processes involved in getting a tax on sugary soft drinks onto the political agenda and into policy and practice in Mexico highlight the importance of coordinated civil-society advocacy in countering the might of transnational food manufacturing giants. It also provides a demonstration of the quantifiable impact of a soda tax on obesity risk and outcomes. A consortium of civil associations, social organizations, and professionals (The Alliance for Healthy Food) concerned about the epidemic of overweight and obesity in Mexico played a key role in mobilizing government commitment to implement a tax on sugar-sweetened beverages in 2014. The Alliance launched a multipronged communications campaign to raise public awareness of the risks of sugary soft drinks, engaged directly with congress members, and entered into dialogue with the Ministry of Finance. This coincided with an opening policy window as Mexico's health minister, and a newly elected president and legislature, supported the tax within a broader fiscal reform agenda (Le Bodo and De Wals, 2017; Roache and Gostin, 2017). The Alliance campaigned for a 20% tax on sugar-sweetened beverages to decrease their consumption. After fierce opposition from the food and beverage industry, Mexico's government passed a 1-peso per liter sales tax on sugar-sweetened beverages that went into effect on the first day of January 2014. This had the effect of increasing their cost by 10%. Evaluation of the effectiveness of the tax found that sugary beverage sales fell an average of 6% (−12 mL/capita/day) in 2014, and that the decline accelerated over time, reaching a 12% drop by December 2014 (Batis et al., 2016).

to think and work in systems, recognizing that health and social inequities and environmental degradation are the emergent properties of complex adaptive systems. Moving away from a pathology focus also means paying attention to and investigating the political determinants of health and the associated policymaking processes, which takes us into the research area of governance (Friel, 2017).

It is impossible to fully understand climate change and health inequities through the prism of any one discipline or methodology. The integration of perspectives and researchers from a range of disciplines, including public health, climate science, systems science, political science, economics, and public administration, is essential. This book is a call for further expansion of the necessary sisterhood between the sciences, social sciences, and humanities. Such an interdisciplinary approach is needed to provide the sort of transformative evidence that is required to address the challenges posed by climate change and health inequity.

Progressive Policy Systems

A systems-based approach is highly suited to engaging with the "wicked" policy problems (Patton, 2011) posed by the interrelationship between climate change, social conditions, and health inequities, and the shared drivers of climate change and health inequities in a global, consumptagenic system. This approach reflects the realities of complex nonlinear dynamics between environmental, social, economic, and health sectors and actors. The design of policy and actions that promote environmental sustainability and equity in health requires an understanding of the feedback processes that take place among many policy domains, including climate adaptation and mitigation, economics, energy, transport, education, and so on. Such complexity should not mean paralysis. Instead, it can help us see the many possible pathways to a shared goal, a society that we want. And as I describe next, we have plenty of evidence concerning what needs to change and what can be done within the system to ensure a fairer, more sustainable, and healthier world.

Policy for What Purpose?

Before detailing what progressive intersectoral policy that addresses the systemic issues highlighted in this book looks like, let us begin by imagining the society that we want, based on a handful of policy principles. Imagine a time when we have macroeconomic policies designed to improve the lot of everyone—economic growth is constrained and becomes a means to an end rather than the end itself. Conditions of life—education,

Box 3.3

Policy Principles for the World We Want

- Macroeconomic policy used to improve the lot of all
- Conditions that support everyone to flourish
- Inclusive societies that welcome difference
- All policies carried out with the lightest of environmental touches

employment, housing, healthcare, elder care—are established to nurture and enable everyone to flourish, regardless of their zip or postal code, sex, or color. There is intolerance of racism and bigotry, and inclusive societies celebrate diversity—a kinder world. And all of this done with the lightest of environmental touches (Friel, 2018) (Box 3.3).

What Does Progressive Policy Look Like?

There is a pressing need for a shift in priorities across a whole number of policy domains if we are to create such a society. There is need and opportunity to reorient economic, social, environmental, and health policies using contemporary understandings of the social causes of health inequity and the causes of climate change.

At the heart of a healthier and environmentally stable world are issues of equality. As noted in the Introduction to this book, global inequality is a huge problem and one that is growing. To take one example, the Australian Council of Social Service (ACOSS) reports that the average income of individuals in the top 20% of income earners in Australia is five times that of individuals in the lowest 20%, while people in the top quintile assessed by wealth are approximately 70 times wealthier than people in the lowest quintile (ACOSS, 2015). Since 1975, the wages of full-time workers in the bottom 10% of income earners in Australia have risen by 15%, from $32,000 to $37,000, while workers in the top 10% have seen their wages increase by 59%, from $65,000 to $103,000 (Leigh, 2013).

Against this background of rising inequality, Wilkinson and Pickett demonstrated in 2010 that more economically equal societies are better for health and well-being. Not only are members of more equal societies physically and psychologically healthier than their counterparts in very unequal societies, but there are fewer social problems such as crime in these societies (Wilkinson, Pickett, and De Vogli, 2010). Evidence has also been accumulating showing that societies with a high degree of equality are better for the environment (Dorling, 2015, 2017; Wilkinson et al., 2010). In these societies, even though overall material well-being may be high, the soaring levels of conspicuous consumption evident in less equal societies are not present. As Dorling (2017) puts it:

> In the more equal rich countries, people on average consume less, produce less waste and emit less carbon [than people in highly unequal affluent countries like the US] . . . almost everything associated with the environment improves when economic equality is greater.

Multiple reasons for this have been posited, with most appearing to flow from the fact that peoples' lives in general are enhanced by living in societies of relative equality, and their heightened well-being produces behavioral changes that are positive from the perspective of climate change and the environment. For example, status anxiety—and thus the pressure to engage in conspicuous consumption to assert and improve one's social status—is high in highly unequal societies and much lower in more equal societies (Dorling, 2017; Wilkinson et al., 2010). Levels of trust are also higher in more equal societies, with the result that there is a willingness to act on behalf of other people and for the public good. This may be another reason accounting for the fact that in more equal societies people in general recycle more, take less frequent flights, consume less water and meat, and produce less waste (Wilkinson et al., 2010). Wilkinson and colleagues (2010) also cite data showing that, compared with less equal countries, business executives in more equal countries are more supportive of government cooperation with international environmental agreements. They strike this positive note:

> Although change will take time, the convenient truth is that greater equality offers not only the possibility of a reduction in consumerism and status competition, but also the development of a more cohesive, social, and sustainable society, which may be essential for our future health and wellbeing. (Wilkinson et al., 2010)

Equality research thus provides powerful support for the view that focus on redistribution rather than on the relentless pursuit of economic growth is a necessary precondition not only for reducing poverty and a whole range of inequities, including the terrible disparities in health outcomes canvassed in this book, but also for effectively responding to climate change and other environmental risks.

In Chapter 1, I cited Butler's claim that humanity currently has enough resources, knowledge, and technology to meet everyone's basic needs while containing climate change within limits that will ensure both its own and the planet's survival (Butler 2003). I pointed out that this is a view shared by the UN Development Program (UNDP), which, in its 2011 Development Report, argued that equitable forms of governance and distribution of resources would produce major development benefits and obviate the need for further fossil fuel–driven economic growth (UNDP, 2011). For a very long time, arguments along the lines advanced by Butler and the UNDP have been marginalized and barely credited, especially within economic policy circles. In light of evidence on the effects of equality accrued by interdisciplinary researchers like Wilkinson, Pickett, De Vogli, and Dorling, this situation is slowly beginning to change. For the moment, the pace of change is slow—but it could well be transformative.

Pursuit of equality requires a major rethinking of global economic systems, including reform of financial systems to better reflect real assets and liabilities; reform of tax systems to tax the imposition of social burdens such as pollution; and more equitably shared resources, income, and work. In his important book on climate change and the health of nations, McMichael (2017) discusses advocacy among some progressive economists for what they describe as a "steady-state" economy in which there is no economic growth. This radical idea responds to the link between economic growth and environmental degradation (Daly, 1974; New Economics Foundation, 2010a, 2010b). It also focuses on the ability to meet the basic needs of all human beings through a fairer distribution of resources—in other words, via redistribution rather than further growth.

There are various policy approaches that can better distribute income and wealth. Economic inequities can be reduced through improved access to quality education and secure and decent employment, and through the provision of living wages or other forms of social protection (Commission on Social Determinants of Health, 2008). The latter must take into account realistic and current costs of healthy living, which requires supportive economic and social policy. These policies must be regularly updated to reflect the costs of adequate nutritious food, shelter, water and sanitation, and social participation (Marmot and Bell, 2012).

We also need investment in natural and social capital commons; more renewable energy policy; promotion of technological innovations that support well-being rather than the pursuit of profit; institutional support for inclusive democratic participation; and the creation of a culture of well-being rather than consumption. Done well, planning and design policies can reduce social and health inequities, help communities adapt to existing climate change, and mitigate against further environmental harm.

In Chapter 2, we saw that the urban heat island effect can be mitigated and the health impacts of climate change in cities managed to some degree by the use of green spaces and effective urban design. But mitigation efforts generally benefit the affluent while failing to assist those most in need. Impoverished households and neighborhoods lack the resources to reduce and adapt to the impacts of climate change (Kovats and Akhtar, 2008; World Health Organization [WHO], 2016d). They cannot afford to invest in healthier homes and neighborhoods, and when their homes are damaged or destroyed as a consequence of climate change, or they suffer injury themselves, they are unlikely to have health or life insurance or insurance for homes and possessions.

Environmentally sustainable development, oriented to health equity, must ensure the supply of basic amenities, including water and housing. Creating more equitable and green housing development requires regulation of land development for urban regeneration, for example, through fair-share housing programs that involve inclusionary zoning and enforcement of green housing laws (National Neighborhood Coalition [NNC], 2001). When planning cities, planners should not design in congestion, air pollution, and ugly roads that waste time and money, cause daily upset, and limit the pleasures of living in cities (WHO, 2016c). When planning urban settlements, planners should not position low-income residential areas far from concentrations of employment. This will save on transportation costs and time and will reduce emissions (Stone, Hess, and Frumkin, 2010). A more equitable distribution of community facilities such as schools, libraries, and clinics needs to be ensured so that residents of lower socioeconomic areas have access to the benefits that urban life can afford.

Addressing urban inequities that will be aggravated by climate change can be difficult because much of the responsibility for creating cities capable of withstanding climate change falls on local governments; how well these governments are resourced varies markedly not only internationally but also within countries

(WHO, 2016c). For example, in the United States in 2015, annual expenditure per citizen by the country's 100 largest cities varied between a low of US $566 for Colorado Springs and a high of US $15,624 for Washington, DC (Ballotpedia, 2015). At the international level, local governments in most low-income countries annually spend on average less than US $20 per citizen (Satterthwaite, 2013).

Countering the health and environmental pressures associated with urban growth requires sustained capital and other forms of investment in rural development as well. Policies aimed at health, sustainable development, and poverty reduction will require action on issues of rural land tenure and rights and rural infrastructure, including health, education, roads, and services. Such policies will also need to support a diversification of and increase in rural employment opportunities (Swaminathan, 2006).

Financial support is fundamental to the required reorientation of policies. After World War II, with many economies overstretched and infrastructure damaged, the Marshall Plan envisaged cross-country structural support to rebuild a prosperous Europe. Given the critical state of humanity as described in this book, a twenty-first-century plan is needed, this time based on principles of equity and sustainability. Developed countries have a special responsibility for achieving these goals given the stark climate injustices, described earlier in the book, and the dominant influence of market-based economic policies that underpin many global health inequities and from which these countries have disproportionately benefited. Low- and middle-income countries are unlikely to be able to provide the funds needed to create healthy living environments while also undertaking climate change adaptation and mitigation actions. A "Green Marshall Plan" for the twenty-first century is needed, therefore, focusing on development that benefits the poorest countries and the poor within all countries. Not without its challenges (Cui et al., 2014), but the Green Climate Fund holds promise. This is one of the funds within the climate finance architecture established under the UN Framework Convention on Climate Change (UNFCCC, 2011). The purpose of the fund is to promote a shift toward low-emission and climate-resilient development pathways in developing countries.

Political Windows of Opportunity and Challenges to Progressive Action

Intersectoral action is clearly key. It is key to climate adaptation implemented in a way that promotes health and social equity; it is key to mitigation of further environmental harm; and it is key to addressing the drivers of consumptagenic systems that fuel climate change and health inequity simultaneously. The development and implementation of such intersectoral policy must be based on principles of equity, or else the status quo of social and health inequities will continue (Commission on Social Determinants of Health, 2008; Whitmee et al., 2015).

Encouragingly, there is increasing recognition among global, national, and local political and policy leaders that environmental harms and social and health inequities are the outcomes of multiple policy interactions and, as such, that responding to them requires policy coherence in the form of complimentary goals and actions across multiple sectors (Macey, 2018; May, Sapotichne, and Workman, 2006; Ruckert et al., 2017; Signal et al., 2013).

This is apparent in the global 2030 Agenda for Sustainable Development (United Nations, 2015). As noted in the Introduction, by explicitly moving away from the notion that any form of development is desirable and recognizing that in order to benefit humanity and improve outcomes for future as well as current generations development must also be sustainable, the Sustainable Development Goals (SDGs) mark a key moment in global policymaking: they ensure that responses to climate change and to environmental issues more generally are no longer siloed away from other major concerns for human progress.

Less constructively, however, the apparent inclusivity of the Agenda disguises a serious failure to address the consumptagenic system that is driving poor health outcomes, health inequities, and climate change. Achievement of the goals is premised on continued global economic growth, and economic growth at a rate of at least 7% of gross domestic product per annum in the least developed countries (SDG 8.1). While the goals recognize the need to "decouple economic growth from environmental degradation" (SDG 8.4), they say very little about how this might be done. Goal 7 focuses on ensuring access to "affordable, reliable, sustainable and modern energy for all," but the specific targets associated with it are very weak. While they promote greater renewable energy use and "cleaner fossil-fuel technology," they do not call for an end to fossil fuel reliance (7.2 and 7.a).

Although appearing to acknowledge the problems associated with growing global inequality, the Agenda's commitment to equality is limited. SDG 10 aims to "reduce inequality within and among countries," but it focuses on income inequality and says nothing about the much greater issue of wealth inequality.

Moreover, it implicitly sanctions an expansion in income inequality (Hickel, 2015). The goal urges only that income growth among the poorest 40% of the population should be achieved and sustained "at a rate higher than the national average" (SDG 10.1).

The commitment in the Agenda for Sustainable Development to reducing poverty is also very weak. The Agenda's first goal is to eradicate extreme poverty by 2030 (SDG 1), but it puts the threshold for extreme poverty at $1.25/day. This figure is contentious, with some scholars arguing that an adequate income to meet a human's basic needs is actually a minimum of $5/day (Hickel, 2015). Adopting a $5/day poverty threshold would mean acknowledging that as many as 60% of the global population lives in poverty (Hickel, 2015). Putting these arguments aside, it is clear that poverty is a tremendous problem globally, and it is not solely a developing-country problem. Based on 2016 data, more than 1 in 10 people in the United States live in poverty. The official poverty rate in the United States is 12.7%, with poverty characterized as an inability to meet basic food costs without spending more than a third of total household income (Center for Poverty Research, 2017).

Prior to the SDGs, it was very encouraging when national-level ministers and heads of state internationally supported WHO's call to action by endorsing the 2008 recommendations of the Commission on the Social Determinants of Health, including—as we saw in the Introduction—to reform social structures that produce inequitable distributions of power, money, and resources (Commission on Social Determinants of Health, 2008). It is good that national and international health policies are increasingly acknowledging the inequities related to social determinants of health and the need for action not just in the health domain but also in other policy domains (Donkin et al., 2018). This is apparent in programs focusing on the "first 1,000 days of life" from conception to the age of 2 years. This period is critical for protecting against malnutrition, particularly stunting, and for promoting health throughout the life course. It requires access to quality health and other social services, including maternal education, provision of adequate nutrition for pregnant and lactating mothers, and support for breastfeeding during a child's first 6 months (FAO, 2017). The sanitation environment is also critical for ensuring that immune status and dietary absorption are not impeded by infection, disease, and other chronic gut problems (Black et al., 2013; Friel and Ford, 2015; Smith and Haddad, 2015). More broadly, the approach of "Health in All Policies" has been endorsed by WHO and the European Union, and implementation has begun in some countries, along with monitoring programs (Baum et al., 2015; Delany et al., 2016; Leppo et al., 2013; Molnar et al., 2016).

There is less to celebrate at the global level in relation to climate change. Although policymakers are now paying attention to intersectoral issues, including agriculture, transport, fuel, buildings, industry, and waste strategies, the global framework for limiting climate change and for adapting to and mitigating its effects has major defects. It is not legally binding, it is not universally supported—most notoriously it has been rejected by the United States—and, as McMichael (2017) points out, even if all voluntary pledges made under the 2015 Paris Agreement are met, an increase in global average temperatures of 3°C by 2100 remains almost certain.

Despite the Trump administration's withdrawal from the Paris Agreement, beacons of hope for climate action can be found at a more local level, with many local governments acting as global vanguards in responding to climate change (WHO, 2016c). Within the United States more than 1,200 states, cities, and business leaders have banded together to continue to work toward meeting the Paris Agreement's targets—networks of hope among institutional resistance. Other developments include the 2014 "Compact of Mayors," under which 206 cities globally have agreed to reduce their GHG emissions and develop mechanisms to limit their vulnerability to climate change. Local governments are coordinating their efforts and sharing knowledge through organizations such as the International Council for Local Environmental Initiatives (ICLEI), which connects more than 1,000 cities and covers around 20% of the world's urban population (WHO, 2016a). Among the array of initiatives adopted, the Colombian capital of Bogotá has closed 97 kilometers of roadway to traffic on Sundays, reducing the city's air pollution and encouraging its inhabitants to use the streets for cycling, walking, and other physical activities (WHO, 2016a).

In Australia, local governments are showing varying levels of sophistication in the ways that they are leading decisions around mitigation of and adaptation to climate change. Many are developing and implementing proactive climate change plans, with an impressive array of policies and programs. Examples of these include a local council initiative which aims to identify and assist vulnerable households to adapt to climate change risks and address rising energy costs. The city of Melbourne has a well-developed suite of policies that include, for example, doubling tree canopy cover for the city's urban forest, upgrading drainage infrastructure, funding more energy-efficient buildings, implementing planning processes to minimize climate risk, and installing various water-sensitive urban-design initiatives.

Governments and local councils are increasingly recognizing that they cannot, and nor should they, do this alone. As noted in Chapter 1, inclusion in the decision-making processes that affect people's lives is one of the three key dimensions of empowerment for health equity. Communities are therefore essential partners when forming fair and equitable responses to climate change and the social determinants of health inequities (Barten et al., 2011). This is because climate change initiatives will be primarily driven from this scale; it is at the level of the local community where the full brunt of climate change and its effects will be faced and risk information will be managed. Box 3.4 provides a brief description of the power of communities in helping to shape local responses to a devastating extreme-weather event, including the role of the local community as a "lay researcher," holding government and other agencies to account through the use of relevant data.

Encountering Resistance

Getting effective action regarding the social determinants of health inequities and effects of climate change remains hugely challenging. Resistance is frequently justified on the basis of the financial costs of providing public goods such as sound urban planning, quality schooling, and universally affordable quality healthcare. Such justifications can be politically potent, even though they often ignore budgetary savings that are likely to accrue over the long term. Even more difficult is getting action on

Box 3.4

Communities Coming Together to Strengthen Health and Social Outcomes in Conditions of Adversity

In 2014, extreme weather conditions in the Victorian region of Gippsland triggered a fire that was heading in the direction of the Hazelwood Power Station and its open cut mine in Morwell. The fire raged out of control for 4 weeks, causing many residents to flee due to the thick blanket of smoke and ash. Levels of small-particle pollution were 10 to 15 times higher than the recommended daily minimum. Essential services such as education, postal delivery, and the courthouse were temporarily suspended, and residents were advised to either stay indoors or find accommodation elsewhere. Residents were not informed of the health risks or preventive measures to take until more than 2 weeks after the fire began. The role of the community has subsequently been on display in response to this particular example of an extreme weather event, the likes of which are predicted to increase in frequency and intensity. A community group—"Voices of the Valley"—was formed and with the assistance of the advocacy network "GetUp" has been campaigning for recognition of the community's plight and to obtain answers to how the events of 2014 occurred. The emergence of this volunteer-led and community-based group is an example of the strength of social capital (Putnam, 1995). Such groups can have broad and positive consequences for the resilience of communities beyond just the issue at hand. The key is to sustain advocacy momentum in order to build a healthy and equitable community for the long term.

Source: Bowen and Friel (2015).

some of the more politically sensitive issues involving the redistribution of power, money, and resources, for example through trade and taxation reform, and through public support for new environmentally sustainable infrastructure (Shankardass et al., 2012).

The current lack of progressive and effective intersectoral policy and action is unsurprising given that focusing on environmental challenges and on those regarding social determinants of health inequity challenges established political and policy assumptions and requires changing the status quo. This process is not straightforward in that some people and institutions benefit from the status quo (Baker et al., 2017; Baum and Friel, 2017). The complexity and boundary-crossing nature of the determinants of climate change and health inequity also make it difficult to allocate responsibility for multiple policy goals. Policy processes reflect the ways in which power is distributed in societies, from the initial stages of getting an issue onto the policy agenda through to policy formation and implementation. While global policymakers and regulators such as the International Monetary Fund (IMF), G20

(Group of Twenty), and the World Bank are now talking about the need to reduce the systemic risks created by the single-minded pursuit of market strategies, little of substance has changed. As Jacobs and Mazzucato (2016) note, "strongly embedded incentives for both asset-holders and senior corporation executives create powerful tendencies towards short-termism in both finance and industry." Given this reality, regulatory bodies may have limited ability to shift their priorities (Braithwaite and Drahos, 2000).

Earlier, I argued that effective intersectoral action to address the climate change–health inequities nexus requires a systems approach. To date, when tackling complex problems such as climate change and health inequities, the tendency has been to simplify the problem and the causal pathways that have produced it and thus to narrow the policy domains that give rise to outcomes of interest. In reality, however, the many variables in the systems that drive health inequity and climate change interact with each other fluidly and often in a nonlinear way. A coherent, integrated approach to policy formulation and implementation in the arenas of climate change and health equity needs to address the many links that operate indirectly and dynamically through environmental, economic, social, and health systems. Combining the many relevant policy domains, and aiming to understand systemic behavior, will help to identify key leverage points or places to intervene most effectively (Ghaffarzadegan, Lyneis, and Richardson, 2011; Richardson, 2011). Until recently, there has been an almost complete lack of analytical tools available to study, clarify, and communicate these linkages. Complex systems thinking, as described here, offers considerable potential to address this gap (Bammer, 2006).

The rapidly burgeoning field of equality research also demonstrates the power of systems thinking. Measures to reduce inequality have been found to produce benefits and to create positive feedback loops that strengthen well-being in societies across a whole range of indicators. The extraordinary power of equality becomes evident through attention to multiple causal connections and by recognizing that certain points of intervention—such as measures to reduce income inequality—create benefits that flow back into the system and are quickly amplified.

Conclusions

It is easy to be overwhelmed with despair given the dreadful state of affairs described in this book. Rather than despair, the book argues for a shift in the way we understand and act to redress climate change and improve global health equity. It demands a change in the status quo. Achieving this will require redressing inequities in power, money, and resources, and in people's daily living conditions. While not straightforward, given that some people benefit from these inequities, the current shifting political sands provide an opportunity to harness global despair, as well as desire and hope for a different society. Collectively, we can change the status quo toward a society where all communities, globally, are able to live with dignity, good health, and in balance with nature (Friel, 2018).

By focusing on the operation of a consumptagenic system that fuels both climate change and health inequities, the book shows where future advocacy and policy action should be directed. Our targets should be the actors, structures, and ideas that embed, facilitate, and normalize the global dominance of a consumptagenic system addicted to growth regardless of the costs. This focus is particularly important in the industrialized food system and in the processes of urbanization.

Thinking about humanity in terms of systems tells us that the status quo described in this book is not static: this means it does not have to be like it is. Some say that recent events, such as the election of Donald Trump as President of the United States, have caused shock waves in global political and social systems (McKee, Greer, and Stuckler, 2017), and others claim that the Earth system has reached a tipping point (Barnosky and Hadly, 2016). Now the systems are working to recalibrate and, as a result, we have an important window of opportunity to shape what the recalibration looks like. We have plenty of evidence concerning what needs to change and what can be done within the system to ensure a fairer, more sustainable, and healthier world. There is no one perfect way. Rather than grabbing at one lever of influence, reducing the existing inequities and achieving positive outcomes depends on networked combinations of different approaches. In a hyperconnected world there are many partners to help create networked combinations of hope.

The Spirit of 1848 Caucus of the American Public Health Association invokes the wisdom of Raymond Williams. During the heyday of anti-nuclear campaigning in the 1970s, Williams emphasized that for the fight to succeed, activists had to be clear about what they were for, not just what they were against. Their role was to "make hope practical, rather than despair convincing" (Williams, 1980). Things can and do change, but change requires a shared vision, a willingness to use different policy and regulatory approaches, courageous leadership, and, frankly, political struggle.

REFERENCES

Afionis, S., Sakai, M., Scott, K., Barrett, J., and Gouldson, A. (2017) "Consumption-based carbon accounting: does it have a future?," *Wiley Interdisciplinary Reviews: Climate Change*, 8(1), pp. e438.

Alessio, R. (2013) "The global crisis of severe acute malnutrition in children," *Lancet*, 382(9908), pp. 1858.

Allan, R. P. (2014) "Climate change: dichotomy of drought and deluge," *Nature Geoscience*, 7(10), pp. 700–701.

Allender, S., Owen, B., Kuhlberg, J., Lowe, J., Nagorcka-Smith, P., Whelan, J., and Bell, C. (2015) "A community based systems diagram of obesity causes," *PLoS One*, 10(7), pp. e0129683.

American Psychological Association (2010) *Psychology and global climate change: addressing a multifacted phenomenon and set of challenges.* Washington, DC: American Psychological Association.

Anand, S. S., Hawkes, C., de Souza, R. J., Mente, A., Dehghan, M., Nugent, R., Zulyniak, M. A., Weis, T., Bernstein, A. M., Krauss, R. M., Kromhout, D., Jenkins, D. J. A., Malik, V., Martinez-Gonzalez, M. A., Mozaffarian, D., Yusuf, S., Willett, W. C., and Popkin, B. M. (2015) "Food consumption and its impact on cardiovascular disease: importance of solutions focused on the globalized food system: a report from the workshop convened by the World Heart Federation," *Journal of the American College of Cardiology*, 66(14), pp. 1590–1614.

Anderson, G. B., and Bell, M. L. (2011) "Heat waves in the United States: mortality risks during heat waves and effect modification by heat wave characteristics in 43 US communities," *Environmental Health Perspectives*, 119, pp. 210–18.

Angkurawaranon, C., Jiraporncharoen, W., Chenthanakij, B., Doyle, P., and Nitsch, D. (2014) "Urbanization and non-communicable disease in

Southeast Asia: a review of current evidence," *Public Health*, 128(10), pp. 886–95.

AP-HealthGAEN (2011) *An Asia Pacific spotlight on health inequity: taking action to address the social and environmental determinants of health inequity in Asia Pacific, 2011*. Canberra: Global Action for Health Equity Network (HealthGAEN).

Atkinson, J. A., Page, A., Wells, R., Milat, A., and Wilson, A. (2015) "A modelling tool for policy analysis to support the design of efficient and effective policy responses for complex public health problems," *Implementation Science*, 10(1), pp. 26–35.

Australian Council of Social Service (ACOSS) (2015) *Inequality Australia*. Available at: https://www.acoss.org.au/inequality/ (Accessed March 20, 2018).

Australian Institute for Health and Welfare (2015) *Life expectancy of Aboriginal and Torres Strait Islander people*. Canberra: Australian Institute for Health and Welfare.

Baffes, J., and Dennis, A. (2013) *Long-term drivers of food prices*. Development Prospects Group.

Baines, J. (2014) "Food price inflation as redistribution: towards a new analysis of corporate power in the world food system," *New Political Economy*, 19(1), pp. 79–112.

Baker, P., and Friel, S. (2014) "Processed foods and the nutrition transition: evidence from Asia," *Obesity Reviews*, 15(7), pp. 564–77.

Baker, P., and Friel, S. (2016) "Transnational food and beverage corporations, ultra-processed food markets and the nutrition transition in Asia," *Globalization and Health*, 12:80, pp. 1–15.

Baker, P., Friel, S., Kay, A., Baum, F., Strazdins, L., and Mackean, T. (2017) "What enables and constrains the inclusion of the social determinants of health inequities in government policy agendas? A narrative review," *International Journal of Health Policy and Management*, 7(2), pp. 101–111.

Baldos, U. L. C., and Hertel, T. W. (2014) "Global food security in 2050: the role of agricultural productivity and climate change," *Australian Journal of Agricultural and Resource Economics*, 58(4), pp. 554–70.

Baldwin, R. E. (2006) "Multilateralising regionalism: spaghetti bowls as building blocs on the path to global free trade," *World Economy*, 29(11), pp. 1451–1518.

Ballotpedia (2015) "Analysis of spending in America's largest cities." Available at: https://ballotpedia.org/Analysis_of_spending_in_America%27s_largest_cities (Accessed January 29, 2018).

Balmford, A., Green, R., and Phalan, B. (2015) "Land for food and land for nature?," *Daedalus*, 144(4), pp. 57–75.

Bammer, G. (2006) "Integration and implementation sciences: building a new specialization," in Perez, P. and Batten, D. (eds.) *Complex science for a complex world: exploring human ecosystems with agents*. Canberra: ANU ePress, pp. 95–111.

Barnett, J. (2011) "Dangerous climate change in the Pacific Islands: food production and food security," *Regional Environmental Change*, 11(1), pp. S229–37.

Barnosky, A. D., and Hadly, E. A. (2016) *Tipping point for planet earth: how close are we to the edge?* London: St. Martin's Press.

Barten, F., Akerman, M., Becker, D., Friel, S., Hancock, T., Mwatsama, M., Rice, M., Sheuya, S., and Stern, R. (2011) "Rights, knowledge, and governance for improved health equity in urban settings," *Journal of Urban Health*, 88(5), pp. 896–905.

Bar-Yam, Y. (2006) "Improving the effectiveness of health care and public health: a multiscale complex systems analysis," *American Journal of Public Health*, 96(3), pp. 459–66.

Batis, C., Rivera, J. A., Popkin, B. M., and Taillie, L. S. (2016) "First-year evaluation of Mexico's tax on nonessential energy-dense foods: an observational study," *PLoS Medicine*, 13(7), pp. e1002057.

Baum, F., and Fisher, M. (2014) "Why behavioural health promotion endures despite its failure to reduce health inequities," *Sociology of Health & Illness*, 36(2), pp. 213–25.

Baum, F., and Friel, S. (2017) "Politics, policies and processes: a multidisciplinary and multimethods research programme on policies on the social determinants of health inequity in Australia," *BMJ Open*, 7(12), pp. e017772.

Baum, F., Graycar, A., Delany-Crowe, T., de Leeuw, E., Bacchi, C., Popay, J., Orchard, L., Colebatch, H., Friel, S., MacDougall, C., Harris, E., Lawless, A., McDermott, D., Fisher, M., Harris, P., Phillips, C., and Fitzgerald, J. (2018) "Understanding Australian policies on public health using social and political science theories: reflections from an Academy of the Social Sciences in Australia Workshop," *Health Promotion International,* day014, doi:10.1093/heapro/day014.

Baum, F., Lawless, A., MacDougall, C., Delany, T., McDermott, D., Harris, E., and Williams, C. (2015) "New norms new policies: Did the Adelaide Thinkers in Residence scheme encourage new thinking about promoting well-being and health in all policies?," *Social Science & Medicine*, 147, pp. 1–9.

Benach, J., Vives, A., Amable, M., Vanroelen, C., Tarafa, G., and Muntaner, C. (2014) "Precarious employment: understanding an emerging social determinant of health," *Annual Review of Public Health*, 35, pp. 229–53.

Bennett, C. M., and Friel, S. (2014) "Impacts of climate change on inequities in child health," *Children (Basel)*, 1(3), pp. 461–73.

Bennett, C. M., and McMichael, A. J. (2010) "Non–heat related impacts of climate change on working populations," *Global Health Action*, 3(1), pp. 5640–50.

Berkman, L. F., and Kawachi, I. (2000) *Social epidemiology*. Oxford, UK: Oxford University Press.

Berry, H. L., Hogan, A., Owen, J., Rickwood, D., and Fragar, L. (2011) "Climate change and farmers' mental health: risks and responses," *Asia Pacific Journal of Public Health*, 23(2 Suppl), pp. 119S–32.

Bezirtzoglou, C., Dekas, K., and Charvalos, E. (2011) "Climate changes, environment and infection: facts, scenarios and growing awareness from the public health community within Europe," *Anaerobe*, 17(6), pp. 337–40.

Bhutta, Z. A., and Salam, R. A. (2012) "Global nutrition epidemiology and trends," *Annals of Nutrition and Metabolism*, 61 (s1), pp. 19–27.

Bhutta, Z. A., Salam, R. A., and Das, J. K. (2013) "Meeting the challenges of micronutrient malnutrition in the developing world," *British Medical Bulletin*, 106, pp. 7–17.

Black, R. E., Allen, L. H., Bhutta, Z. A., Caulfield, L. E., de Onis, M., Ezzati, M., Mathers, C., Rivera, J.; Maternal and Child Undernutrition Study, G. (2008) "Maternal and child undernutrition: global and regional exposures and health consequences," *Lancet*, 371(9608), pp. 243–60.

Black, R. E., Victora, C. G., Walker, S. P., Bhutta, Z. A., Christian, P., de Onis, M., Ezzati, M., Grantham-McGregor, S., Katz, J., Martorell, R., Uauy, R., Maternal and Child Nutrition Study, G. (2013) "Maternal and child undernutrition and overweight in low-income and middle-income countries," *Lancet*, 382(9890), pp. 427–51.

Bourdieu, P. (1989) *Distinction. A social critique of the judgement of taste.* London: Routledge & Kegan Paul.

Bowen, K., and Friel, S. (2015) "Health and social impacts of climate change," in Walker, R., and Mason, W. (eds.) *Climate change adaptation for health and social services.* Victoria: CSIRO Publishing, pp. 3–16.

Bradbear, C., and Friel, S. (2013) "Integrating climate change and health into food policy: an analysis of how climate change can affect food prices and population health," *Food Policy*, 43, pp. 56–66.

Braithwaite, J., and Drahos, P. (2000) *Global business regulation.* New York: Cambridge University Press.

Brinkman, H. J., de Pee, S., Sanogo, I., Subran, L., and Bloem, M. W. (2010) "High food prices and the global financial crisis have reduced access to nutritious food and worsened nutritional status and health," *Journal of Nutrition*, 140(1), pp. 153S–61S.

Bureau of Labor Statistics (2015) "Average household spending for food by household income group, 2015." TED: The Economics Daily. Washington, DC: Bureau of Labor Statistics, U.S. Department of Labor. Available at: https://www.bls.gov/opub/ted/2016/high-income-households-spent-half-of-their-food-budget-on-food-away-from-home-in-2015.htm (Accessed March 29, 2018).

Buse, K., Tanaka, S., and Hawkes, S. (2017) "Healthy people and healthy profits? Elaborating a conceptual framework for governing the commercial determinants of non-communicable diseases and identifying options for reducing risk exposure," *Global Health*, 13(1), pp. 34.

Butler C. (2003). *Ecosystems and human well-being. Millenium Ecosystem Assessment*. Geneva: World Health Organization.

Campbell-Lendrum, D. H., Corvalán, C. F., and Prüss–Ustün, A. (2003) "How much disease could climate change cause?," in *Climate change and human health: risks and responses*. Geneva: World Health Organization.

Caney, S. (2008) "Cosmopolitan justice, responsibility, and global climate change," in Brooks, T. (ed.) *The global justice reader*. Singapore: Blackwell Publishing, pp. 689–713.

Canning, P., Rehkamp, S., Waters, A., and Etemadnia, H. (2017) *The role of fossil fuels in the U.S. food system and the American diet, ERR-224*. Washington, DC: U.S. Department of Agriculture, Economic Research Service.

Carey, G., Malbon, E., Carey, N., Joyce, A., Crammond, B., and Carey, A. (2015) "Systems science and systems thinking for public health: a systematic review of the field," *BMJ Open*, 5(12), pp. e009002.

Catford, J. (2009) "Advancing the 'science of delivery' of health promotion: not just the 'science of discovery'," *Health Promotion International*, 24(1), pp. 1–5.

Cazenave, A., Dieng, H. B., Meyssignac, B., Von Schuckmann, K., Decharme, B., and Berthier, E. (2014) "The rate of sea-level rise," *Nature Climate Change*, 4(5), pp. 358–61.

Center for Poverty Research (2017, December 18) "What is the current poverty rate in the US?—Current estimates on poverty in the United States," Available at: https://poverty.ucdavis.edu/faq/what-current-poverty-rate-united-states.

Chase-Dunn, C., Kawano, Y., and Brewer, B. D. (2000) "Trade globalization since 1795: waves of integration in the world-system," *American Sociological Review*, 65(1), pp. 77–95.

Chen, M., Zhang, H., Liu, W., and Zhang, W. (2014) "The global pattern of urbanization and economic growth: evidence from the last three decades," *PLoS One*, 9(8), pp. e103799.

Chivian, E. (ed.) (2002) *Biodiversity: its importance to human health—interim executive summary*, Cambridge, MA: Center for Health and the Global Environment at Harvard Medical School. Available at: https://www.uttayarndham.org/sites/default/files/3%20Biodiversity_v2_screen_0.pdf.

Chow, W. T. L., Brennan, D., and Brazel, A. J. (2012) "Urban heat island research in Phoenix, Arizona: theoretical contributions and policy applications," *Bulletin of the American Meteorological Society*, 93(4), pp. 517–30.

Clapp, J., Desmarais, A., and Margulis, M. (2015) "State of the world food system," *Canadian Food Studies/La Revue canadienne des études sur l'"alimentation*, 2(2), pp. 7–8.

Clark, B., and York, R. (2005) "Carbon metabolism: global capitalism, climate change, and the biospheric rift," *Theory and Society*, 34(4), pp. 391–428.

Clark, S. E., Hawkes, C., Murphy, S. M., Hansen-Kuhn, K. A., and Wallinga, D. (2012) "Exporting obesity: US farm and trade policy and the transformation of the Mexican consumer food environment," *International Journal of Occupational and Environmental Health*, 18(1), pp. 53–65.

Commission on Social Determinants of Health (2008) *Closing the gap in a generation: health equity through action on the social determinants of health. Final report of the Commission on Social Determinants of Health*. Geneva: World Health Organization.

Corning, P. A. (2002) "The re-emergence of emergence?: a venerable concept in search of a theory," *Complexity*, 7(6), pp. 18–30.

Costa, F., Carvalho-Pereira, T., Begon, M., Riley, L., and Childs, J. (2017) "Zoonotic and vector-borne diseases in urban slums: opportunities for intervention," *Trends in Parasitology*, 33(9), pp. 660–62.

Costello, A., Abbas, M., Allen, A., Ball, S., Bell, S., Bellamy, R., Friel, S., Groce, N., Johnson, A., Kett, M., Lee, M., Legrand, S., Levy, C., Maslin, M., McCoy, D., McGuire, B., Montgomery, H., Napier, D., Pagel, C., Patel, J., Oliveira, J. P. d., Redclift, N., Rees, H., Rogger, D., Scott, J., Smith, S., Stephenson, J., Twigg, J., Wolff, J., and Patterson, C. (2009) "Managing the health effects of climate change. UCL Institute for Global Health and Lancet Commission," *Lancet*, (373), pp. 1693–733.

Cottrill, B., Smith, C., Berry, P., Weightman, R., Wiseman, J., White, G., and Temple, M. (2007) *Opportunities and implications of using the co-products from biofuel production as feeds for livestock*: Report prepared by ADAS and Nottingham University for the Home-Grown Cereals Authority, English Beef and Lamb Executive and British Pig Executive.

Creutzig, F., Baiocchi, G., Bierkandt, R., Pichler, P. P., and Seto, K. C. (2015) "Global typology of urban energy use and potentials for an urbanization mitigation wedge," *Proceedings of the National Academy of Sciences of the United States of America*, 112(20), pp. 6283–8.

Crutzen, P. J. (2002) "The effects of industrial and agricultural practices on atmospheric chemistry and climate during the anthropocene," *Journal of Environmental Science and Health Part A: Toxic/Hazardous Substances & Environmental Engineering*, 37(4), pp. 423–4.

Cui, L. B., Zhu, L., Springmann, M., and Fan, Y. (2014) "Design and analysis of the green climate fund," *Journal of Systems Science and Systems Engineering*, 23(3), pp. 266–99.

Daly, H. E. (1974) "Steady-state economics versus growthmania: a critique of the orthodox conceptions of growth, wants, scarcity, and efficiency," *Policy Sciences*, 5(2), pp. 149–167.

Dawe, D., Morales-Opazo, C., Balie, J., and Pierre, G. (2015) "How much have domestic food prices increased in the new era of higher food prices?," *Global Food Security*, 5, pp. 1–10.

Delany, T., Lawless, A., Baum, F., Popay, J., Jones, L., McDermott, D., Harris, E., Broderick, D., and Marmot, M. (2016) "Health in all policies in South Australia: what has supported early implementation?," *Health Promotion International*, 31(4), pp. 888–98.

de Leeuw, E. (2017) "Engagement of sectors other than health in integrated health governance, policy, and action," *Annual Review of Public Health*, 38, pp. 329–49.

De Schutter, O. (2009) *International trade in agriculture and the right to food*. Geneva World Trade Organization.

Di Cesare, M., Khang, Y. H., Asaria, P., Blakely, T., Cowan, M. J., Farzadfar, F., Guerrero, R., Ikeda, N., Kyobutungi, C., Msyamboza, K. P., Oum, S., Lynch, J. W., Marmot, M. G., Ezzati, M.; for Lancet, N. C. D. Action Group (2013) "Inequalities in non-communicable diseases and effective responses," *Lancet*, 381(9866), pp. 585–97.

Diez Roux, A. V. (2011) "Complex systems thinking and current impasses in health disparities research," *American Journal of Public Health*, 101(9), pp. 1627–34.

Dimitri, C., Effland, A. B., Conklin, N. C., and Dimitri, C. (2005) *The 20th century transformation of US agriculture and farm policy*. Washington, DC: Department of Agriculture, Economic Research Service.

D'Ippoliti, D., Michelozzi, P., Marino, C., de"Donato, F., Menne, B., Katsouyanni, K., Kirchmayer, U., Analitis, A., Medina-Ramon, M., Paldy, A., Atkinson, R., Kovats, S., Bisanti, L., Schneider, A., Lefranc, A., Iniguez, C., and Perucci, C. A. (2010) "The impact of heat waves on mortality in 9 European cities: results from the EuroHEAT project," *Environmental Health*, 9, pp. 37.

Dodman, D., Brown, D., Francis, K., Hardoy, J., Cassidy, J., and Satterthwaite, D. (2013) *Understanding the nature and scale of uban risk in low- and middle income countries and its implications for humanitarian preparedness, planning and response*. London: International Institute for the Environment and Development (IIED).

Donkin, A., Goldblatt, P., Allen, J., Nathanson, V., and Marmot, M. (2018) "Global action on the social determinants of health," *BMJ Global Health*, 3(Suppl 1), pp. e000603.

Dorling, D. (2015) "The mother of underlying causes—economic ranking and health inequality," *Social Science & Medicine*, 128, pp. 327–30.

Dorling, D. (2017) *The equality effect: improving life for everyone*. Oxford, UK: The New Internationalist.

Dowler, E. (2008) "Symposium on 'Intervention policies for deprived households': policy initiatives to address low-income households' nutritional needs in the UK," *Proceedings of the Nutrition Society*, 67(3), pp. 289–300.

Downie, C. (2015) "Global energy governance: do the BRICs have the energy to drive reform?," *International Affairs*, 91(4), pp. 799–812.

Downie, C., and Williams, M. (2018) "After the Paris Agreement: What Role for the BRICS in Global Climate Governance?," *Global Policy*, doi:10.1111/1758-5899.12550

Durkalec, A., Furgal, C., Skinner, M. W., and Sheldon, T. (2015) "Climate change influences on environment as a determinant of Indigenous health: relationships to place, sea ice, and health in an Inuit community," *Social Science & Medicine*, 136, pp. 17–26.

Eckert, S., and Kohler, S. (2014) "Urbanization and health in developing countries: a systematic review," *World Health Population*, 15(1), pp. 7–20.

EcoNexus (2013) "Agropoly—a handful of corporations control world food production." Bern, Switzerland: Berne Declaration & EcoNexus. Available at: https://econexus.info/sites/econexus/files/Agropoly_Econexus_BerneDeclaration.pdf.

Economic and Social Commission for Asia and the Pacific (ESCAP) (2009) *Sustainable agriculture and food security in Asia and the Pacific*. Bangkok: United Nations Economic and Social Commission for Asia and the Pacific.

Egondi, T., Kyobutungi, C., Kovats, S., Muindi, K., Ettarh, R., and Rocklov, J. (2012) "Time-series analysis of weather and mortality patterns in Nairobi's informal settlements," *Global Health Action*, 5, pp. 23–32.

Ellen MacArthur Foundation (2016) *The new plastics economy: rethinking the future of plastics*, Cowes, UK: Ellen MacArthur Foundation.

Esposti, P. D. (2012) "Hyperconsumption," in Ritzer, G. (ed.) *The Wiley-Blackwell encyclopedia of globalization*. Hoboken, NJ: John Wiley & Sons.

Ezeh, A., Oyebode, O., Satterthwaite, D., Chen, Y. F., Ndugwa, R., Sartori, J., Mberu, B., Melendez-Torres, G. J., Haregu, T., Watson, S. I., Caiaffa, W., Capon, A., and Lilford, R. J. (2017) "The history, geography, and sociology of slums and the health problems of people who live in slums," *Lancet*, 389(10068), pp. 547–58.

Ezzati, M., Vander Hoorn, S., Lawes, C. M., Leach, R., James, W. P., Lopez, A. D., Rodgers, A., and Murray, C. J. (2005) "Rethinking the 'diseases of affluence' paradigm: global patterns of nutritional risks in relation to economic development," *PLoS Med*, 2(5), pp. e133.

Fedoroff, N. V., Battisti, D. S., Beachy, R. N., Cooper, P. J., Fischhoff, D. A., Hodges, C. N., Knauf, V. C., Lobell, D., Mazur, B. J., Molden, D., and Reynolds, M. P. (2010) "Radically rethinking agriculture for the 21st century," *Science*, 327(5967), pp. 833–4.

Fink, G., Gunther, I., and Hill, K. (2014) "Slum residence and child health in developing countries," *Demography*, 51(4), pp. 1175–97.

Fischer, G., Tubiello, F. N., Van Velthuizen, H., and Wiberg, D. A. (2007) "Climate change impacts on irrigation water requirements: effects of mitigation, 1990–2080," *Technological Forecasting and Social Change*, 74(7), pp. 1083–107.

Floater, G., Rode, P., Robert, A., Kennedy, C., Hoornweg, D., Slavcheva, R., and Godfrey, N. (2014) "Cities and the new climate economy: the transformative role of global urban growth," New Climate Economy Cities Paper 01. LSE Cities, London School of Economics and Political Science. Available at: https://files.lsecities.net/files/2014/11/LSE-Cities-2014-The-Transformative-Role-of-Global-Urban-Growth-NCE-Paper-01.pdf.

Foley, J. A., Ramankutty, N., Brauman, K. A., Cassidy, E. S., Gerber, J. S., Johnston, M., Mueller, N. D., O"Connell, C., Ray, D. K., West, P. C., Balzer, C., Bennett, E. M., Carpenter, S. R., Hill, J., Monfreda, C., Polasky, S., Rockstrom, J., Sheehan, J., Siebert, S., Tilman, D., and Zaks, D. P. (2011) "Solutions for a cultivated planet," *Nature*, 478(7369), pp. 337–42.

Food and Agriculture Organization (FAO) (2006a) *Livestock's long shadow—environmental issues and options*. Rome: Food and Agriculture Organization of the United Nations.

Food and Agriculture Organization (FAO) (2006b) *World agriculture: towards 2030/2050. Interim report Global Perspective Studies Unit*. Rome: Food and Agriculture Organization of the United Nations.

Food and Agriculture Organization (FAO) (2009) *How to feed the world in 2050*. Rome: Food and Agriculture Organization of the United Nations.

Food and Agriculture Organization (FAO) (2011) *The state of the worlds land and water resources for food and agriculture*. Abingdon: Food and Agriculture Organization of the United Nations with Earthscan.

Food and Agriculture Organization (FAO) (2011) *The State of Food Insecurity in the World: How does international price volatility affect domestic economies and food security?* Rome: Food and Agriculture Organization.

Food and Agriculture Organization (FAO) (2012) *The State of World Fisheries and Aquaculture*. Rome: Food and Agriculture Organization of the United Nations.

Food and Agriculture Organization (FAO) (2017) *The state of food security and nutrition in the world 2017: building resilience for peace and food security*. Rome: Food and Agriculture Organization of the United Nations.

Food Security Information Network (2017) *Global report on food crises 2017.* Rome: World Food Program.

Ford, L., Kirk, M., Glass, K., and Hall, G. (2014) "Sequelae of foodborne illness caused by five pathogens, Australia, circa 2010," *Emerging Infectious Diseases*, 20(11), pp. 1860–66.

Forrester, J. (1961) *Industrial dynamics.* Cambridge, MA: MIT Press.

Forrester, J. (1969) *Urban dynamics.* Cambridge, MA: MIT Press.

Friedlingstein, P., Andrew, R. M., Rogelj, J., Peters, G. P., Canadell, J. G., Knutti, R., Luderer, G., Raupach, M. R., Schaeffer, M., van Vuuren, D. P., and Le Quere, C. (2014) "Persistent growth of CO2 emissions and implications for reaching climate targets," *Nature Geoscience*, 7(10), pp. 709–715.

Friedman, E. A., and Gostin, L. O. (2017) "From local adaptation to activism and global solidarity: framing a research and innovation agenda towards true health equity," *International Journal of Equity in Health*, 16(1), pp. 18.

Friel, S. (2017) "Governance, regulation and health equity," in Drahos, P. (ed.) *Regulatory theory: foundations and applications.* Canberra: ANU ePress, pp. 573–90.

Friel, S. (2018) "A time for hope? Pursuing a vision of a fair, sustainable and healthy world," *Global Policy*, 9(2), pp. 276–282.

Friel, S., and Ford, L. (2015) "Systems, food security and human health," *Food Security*, 7(2), pp. 437–51.

Friel, S., Gleeson, D., Thow, A. M., Labonte, R., Stuckler, D., Kay, A., and Snowdon, W. (2013) "A new generation of trade policy: potential risks to diet-related health from the trans pacific partnership agreement," *Global Health*, 9(46), pp. 1–7.

Friel, S., Hancock, T., Kjellstrom, T., McGranahan, G., Monge, P., and Roy, J. (2011) "Urban health inequities and the added pressure of climate change: an action-oriented research agenda," *Journal of Urban Health*, 88(5), pp. 886–95.

Friel, S., Marmot, M., McMichael, A., Kjellstrom, T., and Vagero, D. (2008) "Global health equity and climate stabilisation: a common agenda," *Lancet*, 372(9650), pp. 1677–83.

Friel, S., Pescud, M., Malbon, E., Lee, A., Carter, R., Greenfield, J., Cobcroft, M., Potter, J., Rychetnik, L., and Meertens, B. (2017) "Using systems science to understand the determinants of inequities in healthy eating," *Plos One*, 12(11), pp. e0188872.

Fussell, E., Sastry, N., and Vanlandingham, M. (2010) "Race, socioeconomic status, and return migration to New Orleans after Hurricane Katrina," *Population and Environment*, 31(1–3), pp. 20–42.

Garnett, T. (2011) "Where are the best opportunities for reducing greenhouse gas emissions in the food system (including the food chain)?," *Food Policy*, 36, pp. S23–S32.

Garnett, T. (2016) "Plating up solutions," *Science*, 353(6305), pp. 1202–4.

Ghaffarzadegan, N., Lyneis, J., and Richardson, G. P. (2011) "How small system dynamics models can help the public policy process," *System Dynamics Review*, 27(1), pp. 22–44.

Ghosh, J. (2010) "The unnatural coupling: food and global finance," *Journal of Agrarian Change*, 10(1), pp. 72–86.

Gibson, R. S. (2011) *Strategies for preventing multi-micronutrient deficiencies: a review of experiences with food-based approaches in developing countries*. Rome: CAB International and FAO.

Glaeser, B. (2010) *The Green Revolution revisited: critique and alternatives*. Taylor & Francis.

Glasgow Centre for Population Health (2014) *Ten years of the Glasgow Centre for Population Health: the evidence and implications*, Glasgow: Glasgow Centre for Population Health. Available at: http://www.understandingglasgow.com/indicators/health/overview.

Goldstein, J. (1999) "Emergence as a construct: history and issues," *Emergence* 11, pp. 49–72.

Goryakin, Y., and Suhrcke, M. (2014) "Economic development, urbanization, technological change and overweight: what do we learn from 244 demographic and health surveys?," *Economics & Human Biology*, 14, pp. 109–27.

Government Office for Science. (2007) *Tackling obesities: future choices—project report*. London: Foresight Programme Government Office for Science.

Grant, W. (1997) *The common agricultural policy*. London: Macmillan.

Green, R., Cornelsen, L., Dangour, A. D., Turner, R., Shankar, B., Mazzocchi, M., and Smith, R. D. (2013) "The effect of rising food prices on food consumption: systematic review with meta-regression," *BMJ*, 346, pp. f3703–11.

Grist, N. (2015) *Topic guide: climate change, food security and agriculture*, London: Evidence on Demand, UK.

Hajat, S., Armstrong, B. G., Gouveia, N., and Wilkinson, P. (2005) "Mortality displacement of heat-related deaths: a comparison of Delhi, Sao Paulo, and London," *Epidemiology*, 16(5), pp. 613–20.

Hanjra, M. A., and Qureshi, M. E. (2010) "Global water crisis and future food security in an era of climate change," *Food Policy*, 35(5), pp. 365–77.

Hanson, L. A., Zahn, E. A., Wild, S. R., Dopfer, D., Scott, J., and Stein, C. (2012) "Estimating global mortality from potentially foodborne diseases: an analysis using vital registration data," *Population Health Metrics*, 10(1), pp. 5.

Harris, G. (2014) "Facing rising seas, Bangladesh confronts the consequences of climate change," *New York Times*, March 28.

Hausfather, Z. (2017) "Mapped: the world's largest CO2 importers and exporters," *CarbonBrief*, July 5. Available at: https://www.carbonbrief.org/mapped-worlds-largest-co2-importers-exporters (Accessed July 2, 2018).

Havinga, T., van Waarden, F., and Casey, D. (2015) *The changing landscape of food governance: public and private encounters*. Cheltenham, UK: Edward Elgar Publishing.

Hawkes, C. (2005) "The role of foreign direct investment in the nutrition transition," *Public Health Nutrition*, 8(4), pp. 357–65.

Hawkes, C. (2006) "Uneven dietary development: linking the policies and processes of globalization with the nutrition transition, obesity and diet-related chronic diseases," *Global Health*, 2(4), pp. 4.

Hawkes, C., Friel, S., Lobstein, T., and Lang, T. (2012) "Linking agricultural policies with obesity and noncommunicable diseases: A new perspective for a globalising world," *Food Policy*, 37(3), pp. 343–53.

Herrero, M., Henderson, B., Havlik, P., Thornton, P. K., Conant, R. T., Smith, P., Wirsenius, S., Hristov, A. N., Gerber, P., Gill, M., Butterbach-Bahl, K., Valin, H., Garnett, T., and Stehfest, E. (2016) "Greenhouse gas mitigation potentials in the livestock sector," *Nature Climate Change*, 6(5), pp. 452–61.

Hickel, J. (2015, September 23) "Five reasons to think twice about the UN's Sustainable Development Goals," *South Asia @ LSE*. Available at: http://blogs.lse.ac.uk/southasia/2015/09/23/five-reasons-to-think-twice-about-the-uns-sustainable-development-goals/ (Accessed July 2, 2018).

Holland, J. H. (2006) "Studying complex adaptive systems," *Journal of Systems Science and Complexity*, 19(1), pp. 1–8.

Hoornweg, D., Sugar, L., and Trejos Gómez, C. L. (2011) "Cities and greenhouse gas emissions: moving forward," *Environment and Urbanization*, 23(1), pp. 207–27.

Horton, R. (2018) "Offline: apostasy against the public health elites," *Lancet*, 391(10121), pp. 643.

Howes, E. L., Joos, F., Eakin, C. M., and Gattuso, J.-P. (2015) "An updated synthesis of the observed and projected impacts of climate change on the chemical, physical and biological processes in the oceans," *Frontiers in Marine Science*, 2(June), pp. 1–27.

Hyatt, O. M., Lemke, B., and Kjellstrom, T. (2010) "Regional maps of occupational heat exposure: past, present, and potential future," *Global Health Action*, 3(1), pp. 5715–25.

Inequality.Org (2018) "Global inequality: facts," Washington, DC: Institute for Policy Studies. Available at: https://inequality.org/facts/global-inequality/ (Accessed: July 2, 2018).

Ingraham, C. (2016) "The stuff we really need is getting more expensive. Other stuff is getting cheaper," *Washington Post*, August 17, 2016. Available at: https://www.washingtonpost.com/news/wonk/wp/2016/08/17/the-stuff-we-really-need-is-getting-more-expensive-other-stuff-is-getting-cheaper/?utm_term=.8746d98d1717.

Intergovernmental Panel on Climate Change (IPCC) (2007) *Climate change 2007: impacts, adaptation and vulnerability. Contribution of Working Group II to the Fourth Assessment Report of the Intergovernmental Panel on Climate Change*. Cambridge, UK, and New York: Cambridge University Press.

Intergovernmental Panel on Climate Change (IPCC) (2014a) *IPCC, 2014: impacts, adaptation, and vulnerability. Part A: global and sectoral aspects. Contribution of Working Group II to the Fifth Assessment Report of the Intergovernmental Panel on Climate Change*. Cambridge, UK, and New York: Cambridge University Press.

Intergovernmental Panel on Climate Change (IPCC) (2014b) "Summary for policymakers," in *Climate change 2014: mitigation of climate change. Contribution of Working Group III to the Fifth Assessment Report of the Intergovernmental Panel on Climate Change.* Cambridge, UK, and New York: Cambridge University Press.

International Energy Agency (2016) "IEA data shows global energy production and consumption continue to rise," Paris, France: International Energy Agency. Available at: https://www.iea.org/newsroom/news/2016/august/iea-data-shows-global-energy-production-and-consumption-continue-to-rise.html (Accessed March 24, 2018).

International Food Policy Research Institute (2015) *Global nutrition report 2015*. Washington, DC: International Food Policy Research Institute.

International Food Policy Research Institute (2017) *Global food policy report 2017*. Washington, DC: International Food Policy Research Institute.

Jabobs, M., and Mazzucato, M. (2016) *Rethinking capitalism: economic policy for sustainable and inclusive growth*. Chichester: Wiley Blackwell.

Jackson, R. B., Canadell, J. G., Le Quere, C., Andrew, R. M., Korsbakken, J. I., Peters, G. P., and Nakicenovic, N. (2015) "Reaching peak emissions," *Nature Climate Change*, 6, pp. 7–10.

James, S. W., and Friel, S. (2015) "An integrated approach to identifying and characterising resilient urban food systems to promote population health in a changing climate," *Public Health Nutrition*, 18(13), pp. 2498–508.

Joffe, M., and Mindell, J. (2006) "Complex causal process diagrams for analyzing the health impacts of policy interventions," *American Journal of Public Health*, 96(3), pp. 473–9.

Johnson, A., and White, N. D. (2014) "Ocean acidification: the other climate change issue," *American Scientist*, 102(1), pp. 60–3.

Jones, A. D., and Ejeta, G. (2016) "A new global agenda for nutrition and health: the importance of agriculture and food systems," *Bulletin of the World Health Organization*, 94(3), pp. 228–9.

Kalkuhl, M., von Braun, J., and Torero, M. (2016) *Food price volatility and its implications for food security and policy*. New York: Springer International Publishing.

Khan, S. A., Aschwanden, A., Bjork, A. A., Wahr, J., Kjeldsen, K. K., and Kjaer, K. H. (2015) "Greenland ice sheet mass balance: a review," *Reports on Progress in Physics*, 78(4), pp. 6801–27.

Khokhar, T. (2017, March 22) "Chart: globally, 70% of freshwater is used for agriculture," *The DATA Blog*. Available at: https://blogs.worldbank.org/opendata/chart-globally-70-freshwater-used-agriculture.

Kissinger, G., Herold, M., and De Sy, V. (2012) *Drivers of deforestation and forest degradation: a synthesis report for REDD+ policy makers*. Vancouver, Canada: Lexeme Consulting.

Kjellstrom, T. (2009a) *Climate change exposures, chronic diseases and mental health in urban populations—a threat to health security, particularly for the poor and disadvantaged. Technical report*. Kobe: World Health Organisation Kobe Centre.

Kjellstrom, T. (2009b) "Climate change, direct heat exposure, health and well-being in low and middle-income countries," *Global Health Action*, 2, pp. 1–3.

Kjellstrom, T., Briggs, D., Freyberg, C., Lemke, B., Otto, M., and Hyatt, O. (2016) "Heat, human performance, and occupational health: a key issue for the assessment of global climate change impacts," *Annual Review of Public Health*, 37(1), pp. 97–112.

Kjellstrom, T., Holmer, I., and Lemke, B. (2009) "Workplace heat stress, health and productivity—an increasing challenge for low and middle-income countries during climate change," *Global Health Action*, 2, pp. 46–51.

Kovats, R. S., and Hajat, S. (2008) "Heat stress and public health: a critical review," *Annual Review of Public Health*, 29, pp. 41–55.

Kovats, S., and Akhtar, R. (2008) "Climate, climate change and human health in Asian cities," *Environment and Urbanization*, 20(1), pp. 165–75.

Krieger, N. (1994) "Epidemiology and the web of causation: has anyone seen the spider?," *Social Science & Medicine*, 39(7), pp. 887–903.

Krieger, N. (2001) "Theories for social epidemiology in the 21st century: an ecosocial perspective," *International Journal of Epidemiology*, 30(4), pp. 668–77.

Krieger, N. (2015) "The real ecological fallacy: epidemiology and global climate change," *Journal of Epidemiology & Community Health*, 69(8), pp. 803–4.

Kyu, H. H., Pinho, C., Wagner, J. A., Brown, J. C., Bertozzi-Villa, A., Charlson, F. J., et al. (2016) "Global and national burden of diseases and injuries among children and adolescents between 1990 and 2013: findings from the Global Burden of Disease 2013 Study," *JAMA Pediatrics*, 170(3), pp. 267–87.

Labonté, R., Schrecker, T., and Gupta, A. (2005) *Health for some: death, disease and disparity in a globalising era*. Toronto: Centre for Social Justice.

Labonté, R., Schrecker, T., Packer, C., and Runnels, V. (2009) *Globalization and health: pathways, evidence and policy*. London: Routledge.

Lam, V. W. Y., Cheung, W. W. L., Swartz, W., and Sumaila, U. R. (2012) "Climate change impacts on fisheries in West Africa: implications for economic, food and nutritional security," *African Journal of Marine Science*, 34(1), pp. 103–17.

Lang, T. (2003) "Food industrialisation and food power: implications for food governance," *Development Policy Review*, 21(5-6), pp. 555–68.

Lang, T., Barling, D., and Caraher, M. (2009) *Food policy: integrating health, environment and society*. Oxford, UK: Oxford University Press.

Lang, T., and Heasman, M. (2004) *Food wars: the global battle for minds, mouths, and markets*. London; Sterling, VA: Earthscan.

Lankao, P. R. (2007) "Are we missing the point? Particularities of urbanization, sustainability and carbon emissions in Latin American cities," *Environment and Urbanization*, 19(1), pp. 159–75.

Larking, E. (2017) "Mobilising for food sovereignty: the pitfalls of international human rights strategies and an exploration of alternatives," *International Journal of Human Rights*, pp. 1–20.

Larking, E. (2018) "Challenging gendered economic and social inequalities: An analysis of the role of trade and financial liberalisation in deepening inequalities, and of the capacity of economic and social rights to redress them," in Harris Rimmer, S. and Ogg, K. (eds.) *The future(s) of feminsit engagement with international law*. Cheltenham, UK: Edward Elgar.

Le Bodo, Y., and De Wals, P. (2017) "Soda taxes: the importance of analysing policy processes; comment on 'The untapped power of soda taxes: incentivising consumers, generating revenue, and altering corporate behaviour'," *International Journal of Health Policy and Management*, 7(5), pp. 470–73.

Lee, J., Schram, A., Riley, E., Harris, P., Baum, F., Fisher, M., Freeman, T., and Friel, S. (2018) "Addressing health equity through action on the social determinants of health: a global review of policy outcome evaluation methods," *International Journal of Health Policy and Management*, 7(7), pp. 581–92.

Leigh, A. (2013) *Battlers & billionaires: the story of inequality in Australia*. Collingwood: Redback.

Leischow, S. J., Best, A., Trochim, W. M., Clark, P. I., Gallagher, R. S., Marcus, S. E., and Matthews, E. (2008) "Systems thinking to improve the public's health," *American Journal of Preventive Medicine* 35(2 Suppl), pp. S196–203.

Leppo, K., Ollila, E., Peña, S., Wismar, M., and Cook, S., Ministry of Social Affairs and Health (2013) *Health in all policies: seizing opportunities, implementing policies*. Helsinki, Finland: Ministry of Social Affairs and Health.

Levin, S., Xepapadeas, T., Crépin, A.-S., Norberg, J., de Zeeuw, A., Folke, C., Hughes, T., Arrow, K., Barrett, S., Daily, G., Ehrlich, P., Kautsky, N., Mäler, K.-G., Polasky, S., Troell, M., Vincent, J. R., and Walker, B. (2012) "Social-ecological systems as complex adaptive systems: modeling and policy implications," *Environment and Development Economics*, 18(2), pp. 111–32.

Lindsay, J. A. (1997) "Chronic sequelae of foodborne disease," *Emerging Infectious Diseases*, 3(4), pp. 443–52.

Lipovetsky, G. (2005) *Hypermodern times*. New York: Polity Press.

Lipovetsky, G. (2011) "The hyperconsumption society," in Ekström, K. M. and Glans, K. (eds.) *Beyond the consumption bubble*, New York: Routledge, pp. 25–36.

Liu, Z., Davis, S. J., Feng, K., Hubacek, K., Liang, S., Anadon, L. D., Chen, B., Liu, J., Yan, J., and Guan, D. (2015) "Targeted opportunities to address the climate–trade dilemma in China," *Nature Climate Change*, 6, pp. 201.

Lobell, D. B., Schlenker, W., and Costa-Roberts, J. (2011) "Climate trends and global crop production since 1980," *Science*, 5, pp. 1204531.

Luber, G., and McGeehin, M. (2008) "Climate change and extreme heat events," *American Journal of Preventive Medicine*, 35(5), pp. 429–35.

Luke, J. H., David, J. F., Erich, M. F., Ed, H., Manoj, J., and Chris, D. J. (2016) "Poorest countries experience earlier anthropogenic emergence of daily temperature extremes," *Environmental Research Letters*, 11(5), pp. 5007.

Lykkeboe, G., and Johansen, K. (1975) "Comparative aspects of buffering capacity in muscle," *Respiratory Physiology*, 25(3), pp. 353–61.

Mabry, P. L., Marcus, S. E., Clark, P. I., Leischow, S. J., and Mendez, D. (2010) "Systems science: a revolution in public health policy research," *American Journal of Public Health*, 100(7), pp. 1161–3.

Macey, A. (2018) "Climate change: towards policy coherence," *Policy Quarterly*, 10(2).

Maniates, M. F. (2010) "Cultivating consumer restraint in an ecologically full world: the case of 'take back your time,'" in Lebel L., Lorek S. and Daniel R. (eds.) *Sustainable production consumption systems*, Dordrecht: Springer, pp. 13–37.

Marcotullio, P. J., Hughes, S., Sarzynski, A., Pincetl, S., Pena, L. S., Romero-Lankao, P., Runfola, D., and Seto, K. C. (2014) "Urbanization and the carbon cycle: contributions from social science," *Earths Future*, 2(10), pp. 496–514.

Marmot, M. (2004) *Status syndrome.* London: Bloomsbury.

Marmot, M., and Bell, R. (2012) "Fair society, healthy lives," *Public Health*, 126 (Suppl 1), pp. S4–10.

Marmot, M., Friel, S., Bell, R., Houweling, T. A. J., Taylor, S., on behalf of the Commission on Social Determinants of Health (2008) "Closing the gap in a generation: health equity through action on the social determinants of health," *Lancet*, 372(9650), pp. 1661–9.

Marmot, M. G., Smith, G. D., Stansfeld, S., Patel, C., North, F., Head, J., White, I., Brunner, E., and Feeney, A. (1991) "Health inequalities among British civil servants: the Whitehall II study," *Lancet*, 337(8754), pp. 1387–93.

Mawani, M., and Aziz Ali, S. (2016) "Iron deficiency anemia among women of reproductive age, an important public health problem: situation analysis," *Reproductive System & Sexual Disorders*, 5(3), pp. 187–93.

May, P. J., Sapotichne, J., and Workman, S. (2006) "Policy coherence and policy domains," *Policy Studies Journal*, 34(3), pp. 381–403.

Mberu, B. U., Haregu, T. N., Kyobutungi, C., and Ezeh, A. C. (2016) "Health and health-related indicators in slum, rural, and urban communities: a comparative analysis," *Global Health Action*, 9, pp. 33163.

McClanahan, T., Allison, E. H., and Cinner, J. E. (2015) "Managing fisheries for human and food security," *Fish and Fisheries*, 16(1), pp. 78–103.

McCorriston, S., Hemming, D., Lamontagne-Godwin, J., Osborn, J., Parr, M., and Roberts, P. (2013) *What is the evidence of the impact of agricultural liberalisation on food security in developing countries? A systematic review*. London: EPPI-Centre, Social Science Research Unit, Institute of Education, University of London.

McCullough, E. B., Pingali, P. L., Stamoulis, K. G., and Food Agriculture Organization of the United Nations (2008) *The transformation of agri-food systems: globalization, supply chains and smallholder farmers*. Rome; London; Sterling, VA: Food and Agriculture Organization of the United Nations; Earthscan.

McGranahan, G., and Satterthwaite, D. (2014) "Urbanisation: concepts and trends," IIED. Available at: https://www.environmentandurbanization.org/urbanisation-concepts-and-trends.

McKee, M., Greer, S. L., and Stuckler, D. (2017) "What will Donald Trump's presidency mean for health? A scorecard," *Lancet*, 389(10070), pp. 748–54.

McLaughlin, K. A., Fairbank, J. A., Gruber, M. J., Jones, R. T., Lakoma, M. D., Pfefferbaum, B., Sampson, N. A., and Kessler, R. C. (2009) "Serious emotional disturbance among youths exposed to Hurricane Katrina 2 years postdisaster," *Journal of the American Academy of Child and Adolescent Psychiatry*, 48(11), pp. 1069–78.

McMichael, A. J. (1994) *Planetary overload: global environmental change and the health of the human species*. Cambridge, UK: Cambridge University Press.

McMichael, A. J. (1999) "Prisoners of the proximate: loosening the constraints on epidemiology in an age of change," *American Journal of Epidemiology*, 149(10), pp. 887–97.

McMichael, A. J. (2009) "Climate change and human health," *Commonwealth Health Ministers' update 2009*. Woodridge, UK: Pro-Book Publishing, pp. 12–21.

McMichael, A. J. (2017) *Climate change and the health of nations: famines, fevers, and the fate of populations*. Oxford, UK: Oxford University Press.

McMichael, A. J., and Beaglehole, R. (2000) "The changing global context of public health," *Lancet*, 356(9228), pp. 495–9.

McMichael, A. J., Butler, C. D., and Dixon, J. (2015) "Climate change, food systems and population health risks in their eco-social context," *Public Health*, 129(10), pp. 1361–8.

McMichael, C. (2016) "Winds of change: climate change, migration and health," in Thomas, F. (ed.) *Handbook of migration and health*, Cheltenham, UK: Edward Elgar, pp. 277–96.

McMichael, C., Barnett, J., and McMichael, A. J. (2012) "An ill wind? Climate change, migration, and health," *Environmental Health Perspectives*, 120(5), pp. 646–54.

McMichael, P. (1994) *The global restructuring of agro-food systems*. Ithaca, NY: Cornell University Press.

McMichael, P. (2009) "A food regime analysis of the 'world food crisis,'" *Agriculture and Human Values*, 26(4), pp. 281–95.

Meadows, D. (2008) *Thinking in systems: a primer*. Vermont: Chelsea Green.

Mendler de Suarez, J., Cicin-Sain, B., Wowk, K., Payet, R., and Hoegh-Guldberg, O. (2014) "Ensuring survival: oceans, climate and security," *Ocean & Coastal Management*, 90, pp. 27–37.

Midgley, G. (2006) "Systemic intervention for public health," *American Journal of Public Health*, 96(3), pp. 466–72.

Milstein, B., Jones, A., Homer, J. B., Murphy, D., Essien, J., and Seville, D. (2007) "Charting plausible futures for diabetes prevalence in the United States: a role for system dynamics simulation modeling," *Preventing Chronic Disease*, 4(3), pp. A52.

Molnar, A., Renahy, E., O"Campo, P., Muntaner, C., Freiler, A., and Shankardass, K. (2016) "Using win-win strategies to implement health in all policies: a cross-case analysis," *PLoS One*, 11(2), pp. e0147003.

Monteiro, C. A., Levy, R. B., Claro, R. M., de Castro, I. R., and Cannon, G. (2011) "Increasing consumption of ultra-processed foods and likely impact on human health: evidence from Brazil," *Public Health Nutrition*, 14(1), pp. 5–13.

Moodie, R., Stuckler, D., Monteiro, C., Sheron, N., Neal, B., Thamarangsi, T., Lincoln, P., Casswell, S., and Lancet, N. C. D. A. G. (2013) "Profits and pandemics: prevention of harmful effects of tobacco, alcohol, and ultra-processed food and drink industries," *Lancet*, 381(9867), pp. 670–9.

Motesharrei, S., Rivas, J., Kalnay, E., Asrar, G. R., Busalacchi, A. J., Cahalan, R. F., Cane, M. A., Colwell, R. R., Feng, K. S., Franklin, R. S., Hubacek, K., Miralles-Wilhelm, F., Miyoshi, T., Ruth, M., Sagdeev, R., Shirmohammadi, A., Shukla, J., Srebric, J., Yakovenko, V. M., and Zeng, N. (2016) "Modeling sustainability: population, inequality, consumption, and bidirectional coupling of the Earth and human systems," *National Science Review*, 3(4), pp. 470–94.

Murray, N. E., Quam, M. B., and Wilder-Smith, A. (2013) "Epidemiology of dengue: past, present and future prospects," *Clinical Epidemiology*, 5, pp. 299–309.

Myers, S. S., Smith, M. R., Guth, S., Golden, C. D., Vaitla, B., Mueller, N. D., Dangour, A. D., and Huybers, P. (2017) "Climate change and global food systems: potential impacts on food security and undernutrition," *Annual Review of Public Health*, 38(1), pp. 259–77.

National Neighborhood Coalition (NNC) (2001) *Smart growth for neighborhoods: affordable housing and regional vision*. Washington, DC: National Neighborhood Coalition.

National Oceanic and Atmospheric Administration (NOAA) (2016) "Global climate report—June 2016," Available at: https://www.ncdc.noaa.gov/sotc/global/201606 (Accessed August 4, 2016).

National Oceanic and Atmospheric Administration (NOAA) (2017) "What is ocean acidification?" Available at: https://www.pmel.noaa.gov/co2/story/What+is+Ocean+Acidification%3F (Accessed July 2, 2018).

Navarro, V. (2000) *The political economy of social inequalities: consequences for health and quality of life*. Amityville, NY: Baywood Publishing Company.

Neff, R. A., Parker, C. L., Kirschenmann, F. L., Tinch, J., and Lawrence, R. S. (2011) "Peak oil, food systems, and public health," *American Journal of Public Health*, 101(9), pp. 1587–97.

Nelson, G. C., Rosegrant, M. W., Koo, J., Robertson, R., Sulser, T., Zhu, T., Ringler, C., Msangi, S., Palazzo, A., Batka, M., Magalhaes, M., Valmonte-Santos, R., Ewing, M., and Lee, D. (2009) *Climate change: impact on agriculture and costs of adaptation*. Washington DC: International Food Policy Research Institute.

Neumann, B., Vafeidis, A. T., Zimmermann, J., and Nicholls, R. J. (2015) "Future coastal population growth and exposure to sea-level rise and coastal flooding—a global assessment," *PLoS One*, 10(3), pp. e0118571.

New Economics Foundation. (2010a) *The great transition: how it all turned out right*, London: New Economics Foundation.

New Economics Foundation. (2010b) *Growth isn't possible: why we need a new economic direction*. London: New Economic Foundation.

Newell, B., Proust, K., Dyball, R., and Mcmanus, P. (2007) "Seeing obesity as a systems problem," *New South Wales Public Health Bulletin*, 18(12), pp. 214–18.

Ng, M., Fleming, T., Robinson, M., Thomson, B., Graetz, N., Margono, C., et al. (2014) "Global, regional, and national prevalence of overweight and obesity in children and adults during 1980–2013: a systematic analysis for the Global Burden of Disease Study 2013," *Lancet*, 384(9945), pp. 766–81.

Nicholls, S. G., Gwozdz, W., Reisch, L., and Voigt, K. (2011) "Fiscal food policy: equity and practice," *Perspectives in Public Health*, 131(4), pp. 157–8.

Oxfam Australia (2018) "Glossary". Food 4 Thought—For students! Available at: https://www.oxfam.org.au/get-involved/how-schools-can-get-involved/classroom-resources/food-4-thought-2/ (Accessed July 2, 2018).

Pachauri, R. K., Allen, M. R., Barros, V. R., Broome, J., Cramer, W., Christ, R., Church, J. A., Clarke, L., Dahe, Q., and Dasgupta, P. (2014) *IPCC, 2014: climate change 2014: synthesis report. Contribution of Working Groups I, II and III to the Fifth Assessment Report of the Intergovernmental Panel on Climate Change*. Geneva: IPCC.

Pataki, D. E., Carreiro, M. M., Cherrier, J., Grulke, N. E., Jennings, V., Pincetl, S., Pouyat, R. V., Whitlow, T. H., and Zipperer, W. C. (2011) "Coupling biogeochemical cycles in urban environments: ecosystem services, green solutions, and misconceptions," *Frontiers in Ecology and the Environment*, 9(1), pp. 27–36.

Patel, R. C. (2012) *Stuffed and starved: the hidden battle for the world food system*. Westminster: Melville House Publishing.

Patton, M. Q. (2011) *Developmental evaluation: applying complexity concepts to enhance innovation and use*. New York: Guilford Press.

Patz, J. A., Gibbs, H. K., Foley, J. A., Rogers, J. V., and Smith, K. R. (2007) "Climate change and global health: Quantifying a growing ethical crisis," *EcoHealth*, 4(4), pp. 397–405.

Patz, J. A., Graczyk, T. K., Geller, N., and Vittor, A. Y. (2000) "Effects of environmental change on emerging parasitic diseases," *International Journal of Parasitology*, 30(12–13), pp. 1395–405.

Patz, J. A., and Olson, S. H. (2006) "Climate change and health: global to local influences on disease risk," *Annals of Tropical Medicine & Parasitology*, 100(5–6), pp. 535–49.

Pelletier, N., Audsley, E., Brodt, S., Garnett, T., Henriksson, P., Kendall, A., Kramer, K. J., Murphy, D., Nemecek, T., and Troell, M. (2011) "Energy intensity of agriculture and food systems," *Annual Review of Environment and Resources*, 36(1), pp. 223–46.

Pfeiffer, D. (2009) *Eating fossil fuels: oil, food and the coming crisis in agriculture*. Gabriola Island, BC: New Society Publishers.

Piesse, J., and Thirtle, C. (2009) "Three Bubbles and a Panic: An Explanatory Review of Recent Food Commodity Price Events," *Food Policy*, 34, 119–29.

Pinard, C. A., Kim, S. A., Story, M., and Yaroch, A. L. (2013) "The food and water system: impacts on obesity," *Journal of Law & Medical Ethics*, 41(Suppl 2), pp. 52–60.

Pinstrup-Andersen, P. (2013) "Nutrition-sensitive food systems: from rhetoric to action," *Lancet*, 382(9890), pp. 375–76.

Polanyi, K., and MacIver, R. M. (1944) *The great transformation*. Boston: Beacon Press.

Popay, J., Escorel, S., Hernandez, M., Johnston, H., Mathieson, J., and Rispel, L. (2008) *Understanding and tackling social exclusion: final report to the WHO Commission on Social Determinants of Health From the Social Exclusion Knowledge Network*. Geneva: World Health Organization.

Popkin, B. M. (2017) "Relationship between shifts in food system dynamics and acceleration of the global nutrition transition," *Nutrition Reviews*, 75(2), pp. 73–82.

Popkin, B. M., Adair, L. S., and Ng, S. W. (2012) "Global nutrition transition and the pandemic of obesity in developing countries," *Nutrition Reviews*, 70(1), pp. 3–21.

Pretty, J. N., Ball, A. S., Lang, T., and Morison, J. I. L. (2005) "Farm costs and food miles: an assessment of the full cost of the UK weekly food basket," *Food Policy*, 30(1), pp. 1–19.

Proust, K., Newell, B., Brown, H., Capon, A., Browne, C., Burton, A., Dixon, J., Mu, L., and Zarafu, M. (2012) "Human health and climate change: leverage points for adaptation in urban environments," *International Journal of Environmental Research and Public Health*, 9(6), pp. 2134–58.

Putnam, R. D. (1995) "Bowling alone: America's declining social capital," *Journal of Democracy*, 6(65–78).

Quaschning, V. (2016) "Statistics: development of global carbon dioxide emissions and concentration in atmosphere." Renewable energy and climate change. Available at: https://www.volker-quaschning.de/datserv/CO2/index_e.php (Accessed July 2, 2018).

Raphael, D. (2015) "Beyond policy analysis: the raw politics behind opposition to healthy public policy," *Health Promotion International*, 30(2), pp. 380–96.

Reckien, D., Creutzig, F., Fernandez, B., Lwasa, S., Tovar-Restrepo, M., McEvoy, D., and Satterthwaite, D. (2017) "Climate change, equity and the Sustainable Development Goals: an urban perspective," *Environment and Urbanization*, 29(1), pp. 159–82.

Reubi, D., Herrick, C., and Brown, T. (2016) "The politics of non-communicable diseases in the global South," *Health Place*, 39, pp. 179–87.

Revi, A. (2008) "Climate change risk: an adaptation and mitigation agenda for Indian cities," *Environment and Urbanization*, 20(1), pp. 207–29.

Richardson, G. P. (2011) "Reflections on the foundations of system dynamics," *System Dynamics Review*, 27(3), pp. 219–43.

Rickles, D., Hawe, P., and Shiell, A. (2007) "A simple guide to chaos and complexity," *Journal of Epidemiology and Community Health* 61(11), pp. 933–7.

Roache, S. A., and Gostin, L. O. (2017) "The untapped power of soda taxes: incentivizing consumers, generating revenue, and altering corporate behavior," *International Journal of Health Policy Management*, 6(9), pp. 489–93.

Robine, J. M., Cheung, S. L., Le Roy, S., Van Oyen, H., Griffiths, C., Michel, J. P., and Herrmann, F. R. (2008) "Death toll exceeded 70,000 in Europe during the summer of 2003," *Comptes Rendus Biologies*, 331(2), pp. 171–8.

Rockstrom, J., Steffen, W., Noone, K., Persson, A., Chapin, F. S., 3rd, Lambin, E. F., Lenton, T. M., Scheffer, M., Folke, C., Schellnhuber, H. J., Nykvist, B., de Wit, C. A., Hughes, T., van der Leeuw, S., Rodhe, H., Sorlin, S., Snyder, P. K., Costanza, R., Svedin, U., Falkenmark, M., Karlberg, L., Corell, R. W., Fabry, V. J., Hansen, J., Walker, B., Liverman, D., Richardson, K., Crutzen, P., and Foley, J. A. (2009) "A safe operating space for humanity," *Nature*, 461(7263), pp. 472–5.

Roser, M., and Ritchie, H. (2018) "Food prices." Available at: https://ourworldindata.org/food-prices (Accessed March 29, 2018).

Ruckert, A., Schram, A., Labonte, R., Friel, S., Gleeson, D., and Thow, A. M. (2017) "Policy coherence, health and the sustainable development goals: a health impact assessment of the Trans-Pacific Partnership," *Critical Public Health*, 27(1), pp. 86–96.

Rutter, H., Savona, N., Glonti, K., Bibby, J., Cummins, S., Finegood, D. T., Greaves, F., Harper, L., Hawe, P., Moore, L., Petticrew, M., Rehfuess, E., Shiell, A., Thomas, J., and White, M. (2017) "The need for a complex systems model of evidence for public health," *Lancet*, 390(10112), pp. 2602–2604.

Rydin, Y., Bleahu, A., Davies, M., Davila, J. D., Friel, S., De Grandis, G., Groce, N., Hallal, P. C., Hamilton, I., Howden-Chapman, P., Lai, K. M., Lim, C. J., Martins, J., Osrin, D., Ridley, I., Scott, I., Taylor, M., Wilkinson, P., and Wilson, J. (2012) "Shaping cities for health: complexity and the planning of urban environments in the 21st century," *Lancet*, 379(9831), pp. 2079–108.

Sage, C. (2013) "The interconnected challenges for food security from a food regimes perspective: energy, climate and malconsumption," *Journal of Rural Studies*, 29, pp. 71–80.

Sanders, R. (2015) "No cause for Caribbean to celebrate Paris Climate Agreement," Antillean Media Group. Available from: https://www.antillean.org/paris-no-celebration-213/.

Sanson-Fisher, R. W., D"Este, C. A., Carey, M. L., Noble, N., and Paul, C. L. (2014) "Evaluation of systems-oriented public health interventions: alternative research designs," *Annual Review of Public Health*, 35(1), pp. 9–27.

Satterthwaite, D. (2013) "The political underpinnings of cities' accumulated resilience to climate change," *Environment and Urbanization*, 25(2), pp. 381–91.

Satterthwaite, D., Huq, S., Pelling, M., and Reid, H. (2007) *Adapting to climate change in urban areas—the possibilities and constraints in low- and middle-income nations*. London: International Institute for Environment and Development.

Savaresi, A. (2016) "The Paris Agreement: a new beginning?," *Journal of Energy & Natural Resources Law*, 34(1), pp. 16–26.

Scallan, E., Hoekstra, R. M., Angulo, F. J., Tauxe, R. V., Widdowson, M. A., Roy, S. L., Jones, J. L., and Griffin, P. M. (2011) "Foodborne illness acquired in the United States—major pathogens," *Emerging Infectious Diseases*, 17(1), pp. 7–15.

Scambler, G. (2007) "Social structure and the production, reproduction and durability of health inequalities," *Social Theory & Health*, 5(4), pp. 297–315.

Schram, A., Friel, S., Freeman, T., Fisher, M., Baum, F., and Harris, P. (2018) "Digital infrastructure as a determinant of health equity: an Australian case study of the implementation of the National Broadband Network," *Australian Journal of Public Administration*, April, doi: 10.1111/1467-8500.12323.

Schram, A., Labonte, R., Baker, P., Friel, S., Reeves, A., and Stuckler, D. (2015) "The role of trade and investment liberalization in the sugar-sweetened carbonated beverages market: a natural experiment contrasting Vietnam and the Philippines," *Global Health*, 11(1), pp. 41.

Sen, A. (1999) *Development as freedom*. New York: Alfred A. Knopf.

Seto K. C., Dhakal, S. Bigio, A., Blanco, H., Delgado, G. C., Dewar, D., Huang, L., Inaba, A., Kansal, A., Lwasa, S., McMahon, J. E., Müller, D. B., Murakami, J., Nagendra H., and Ramaswami, A. (2014) "Human settlements, infrastructure, and spatial planning," in Edenhofer, O., Pichs-Madruga, R. Sokona, Y. Farahani, E. Kadner, S. Seyboth, K. Adler, A. Baum, I. Brunner, S. Eickemeier, P. Kriemann, B. Savolainen, J. Schlömer, S. von Stechow, C. Zwickel, T., and Minx, J. C. (eds.) *Climate change 2014: mitigation of climate change: contribution of Working Group III to the Fifth Assessment Report of the Intergovernmental Panel on Climate Change*. Cambridge, UK; New York: Cambridge University Press, pp. 923–1000.

Shankardass, K., Solar, O., Murphy, K., Greaves, L., and O'Campo, P. (2012) "A scoping review of intersectoral action for health equity involving governments," *International Journal of Public Health*, 57(1), pp. 25–33.

Sharkey, P. (2007) "Survival and death in New Orleans—an empirical look at the human impact of Katrina," *Journal of Black Studies*, 37(4), pp. 482–501.

Short, E. E., Caminade, C., and Thomas, B. N. (2017) "Climate change contribution to the emergence or re-emergence of parasitic diseases," *Infectious Diseases: Research and Treatment*, 10. Available at: http://journals.sagepub.com/doi/full/10.1177/1178633617732296.

Shue, H. (1992) "The unavoidability of justice," in Hurrell, A. and Kingsbury, B. (eds.) *The international politics of the environment*, Oxford, UK: Clarendon Press, pp. 373–97.

Signal, L. N., Walton, M. D., Mhurchu, C. N., Maddison, R., Bowers, S. G., Carter, K. N., Gorton, D., Heta, C., Lanumata, T. S., and McKerchar, C. W.

(2013) "Tackling 'wicked' health promotion problems: a New Zealand case study," *Health Promotion International*, 28(1), pp. 84–94.

Simms, A., and Johnson, V. (2010) *Growth isn"t possible: Why we need a new economic direction*. London: New Economics Foundation.

Singer, P. (2002) *One world: the ethics of globalisation*. Melbourne: Text Publishing.

Smith, K. R. (2008) "Mitigating, adapting, and suffering: how much of each?," *Annual Review of Public Health*, 29(xxxii).

Smith, L. C., and Haddad, L. (2015) "Reducing child undernutrition: past drivers and priorities for the post-MDG era," *World Development*, 68, pp. 180–204.

Son, J. Y., Lee, J. T., Anderson, G. B., and Bell, M. L. (2012) "The impact of heat waves on mortality in seven major cities in Korea," *Environmental Health Perspectives*, 120(4), pp. 566–71.

Spinoni, J., Naumann, G., Carrao, H., Barbosa, P., and Vogt, J. (2014) "World drought frequency, duration, and severity for 1951–2010," *International Journal of Climatology*, 34(8), pp. 2792–804.

Springmann, M., Mason-D"Croz, D., Robinson, S., Garnett, T., Godfray, H. C., Gollin, D., Rayner, M., Ballon, P., and Scarborough, P. (2016) "Global and regional health effects of future food production under climate change: a modelling study," *Lancet*, 387(10031), pp. 1937–46.

Steffen, W., Crutzen, J., and McNeill, J. R. (2007) "The anthropocene: are humans now overwhelming the great forces of nature?," *Ambio*, 36(8), pp. 614–21.

Steffen, W., Richardson, K., Rockstrom, J., Cornell, S. E., Fetzer, I., Bennett, E. M., Biggs, R., Carpenter, S. R., de Vries, W., de Wit, C. A., Folke, C., Gerten, D., Heinke, J., Mace, G. M., Persson, L. M., Ramanathan, V., Reyers, B., and Sorlin, S. (2015) "Sustainability. Planetary boundaries: guiding human development on a changing planet," *Science*, 347(6223), pp. 1259855.

Sterman, J. D. (2000) *Business dynamics: systems thinking and modeling for a complex world*. New York: McGraw-Hill.

Sterman, J. D. (2006) "Learning from evidence in a complex world," *Am J Public Health*, 96(3), pp. 505–14.

Stone, B., Hess, J. J., and Frumkin, H. (2010) "Urban form and extreme heat events: are sprawling cities more vulnerable to climate change than compact cities?," *Environmental Health Perspectives*, 118(10), pp. 1425–8.

Strauss, B., Kulp, S., and Levermann, A. (2015) *Mapping choices: carbon, climate, and rising seas—our global legacy*. Princeton, NJ: Climate Central.

Stuckler, D., and Nestle, M. (2012) "Big food, food systems, and global health," *PLoS Medicine*, 9(6), pp. e1001242.

Susser, M., and Susser, E. (1996) "Choosing a future for epidemiology: II. From black box to Chinese boxes and eco-epidemiology," *American Journal of Public Health*, 86(5), pp. 674–7.

Swaminathan, M. (2006) "2006–7 year of agricultural renewal". Presented at the 93rd Indian Science Congress, Hyderabad.

Swinburn, B. A., Sacks, G., Hall, K. D., McPherson, K., Finegood, D. T., Moodie, M. L., and Gortmaker, S. L. (2011) "The global obesity pandemic: shaped by global drivers and local environments," *Lancet*, 378(9793), pp. 804–14.

Tacoli, C., McGranahan, G., and Satterthwaite, D. (2015) *Urbanisation, rural-urban migration and urban poverty*. Lonodon: IIED.

Thow, A. M., Heywood, P., Schultz, J., Quested, C., Jan, S., and Colagiuri, S. (2011) "Trade and the nutrition transition: strengthening policy for health in the Pacific," *Ecology of Food and Nutrition*, 50(1), pp. 18–42.

Thow, A. M., Jan, S., Leeder, S., and Swinburn, B. (2010) "The effect of fiscal policy on diet, obesity and chronic disease: a systematic review," *Bulletin of the World Health Organization*, 88(8), pp. 609–14.

Thow, A. M., Snowdon, W., Labonte, R., Gleeson, D., Stuckler, D., Hattersley, L., Schram, A., Kay, A., and Friel, S. (2015) "Will the next generation of preferential trade and investment agreements undermine prevention of noncommunicable diseases? A prospective policy analysis of the Trans Pacific Partnership Agreement," *Health Policy*, 119(1), pp. 88–96.

Titley, D. (2017) "Why is climate change's 2 degrees Celsius of warming limit so important?," *The Conversation*. Available at: http://theconversation.com/why-is-climate-changes-2-degrees-celsius-of-warming-limit-so-important-82058.

Troeger, C., Forouzanfar, M., Rao, P. C., Khalil, I., Brown, A., Reiner, R. C., Fullman, N., Thompson, R. L., Abajobir, A., Ahmed, M., Alemayohu, M. A., Alvis-Guzman, N., Amare, A. T., Antonio, C. A., Asayesh, H., Avokpaho, E., Awasthi, A., Bacha, U., Barac, A., Betsue, B. D., Beyene, A. S., Boneya, D. J., Malta, D. C., Dandona, L., Dandona, R., Dubey, M., Eshrati, B., Fitchett, J. R. A., Gebrehiwot, T. T., Hailu, G. B., Horino, M., Hotez, P. J., Jibat, T., Jonas, J. B., Kasaeian, A., Kissoon, N., Kotloff, K., Koyanagi, A., Kumar, G. A., Rai, R. K., Lal, A., El Razek, H. M. A., Mengistie, M. A., Moe, C., Patton, G., Platts-Mills, J. A., Qorbani, M., Ram, U., Roba, H. S., Sanabria, J., Sartorius, B., Sawhney, M., Shigematsu, M., Sreeramareddy, C., Swaminathan, S., Tedla, B. A., Jagiellonian, R. T. M., Ukwaja, K., Werdecker, A., Widdowson, M. A., Yonemoto, N., Zaki, M. E., Lim, S. S., Naghavi, M., Vos, T., Hay, S. I., Murray, C. J. L., and Mokdad, A. H. (2017) "Estimates of global, regional, and national morbidity, mortality, and aetiologies of diarrhoeal diseases: a systematic analysis for the Global Burden of Disease Study 2015," *Lancet Infectious Diseases*, 17(9), pp. 909–48.

Turok, I., and McGranahan, G. (2013) "Urbanization and economic growth: the arguments and evidence for Africa and Asia," *Environment and Urbanization*, 25(2), pp. 465–82.

Uejio, C. K., Wilhelmi, O. V., Golden, J. S., Mills, D. M., Gulino, S. P., and Samenow, J. P. (2011) "Intra-urban societal vulnerability to extreme heat: the role of heat exposure and the built environment, socioeconomics, and neighborhood stability," *Health Place*, 17(2), pp. 498–507.

UN-Habitat (2008) *State of the world cities. Harmonious Cities 2008/2009.* Nairobi: United Nations Human Settlements Program.

UN-Habitat (2016) *Urbanization and development: emerging futures. World Cities Report 2016.* Nairobi: United Nations Human Settlements Program.

United Nations (UN) (2015) *Transforming our world: the 2030 agenda for sustainable development* (A/RES/70/1). New York: United Nations. Available at: https://sustainabledevelopment.un.org/content/documents/21252030%20Agenda%20for%20Sustainable%20Development%20web.pdf.

United Nations (UN) (2018) *United Nations Sustainable Development Goals, 17 goals to transform our world.*

United Nations Department of Economic and Social Affairs (UNDESA) (2011) *World Economic and Social Survey 2011: the great green technological transformation* (E/2011/50/Rev.1 ST/ESA/333). New York: United Nations Department of Economic and Social Affairs.

United Nations Development Programme (UNDP) (2011) *Sustainability and equity: A better future for all.* New York: UNDP.

United Nations Framework Convention on Climate Change (UNFCCC) (2011) *Report of the Conference of the Parties on its sixteenth session. Addendum. Part two. FCCC/CP/2010/7/Add.1.* Bonn: United Nations Framework Convention on Climate Change.

United Nations Framework Convention on Climate Change (UNFCCC) (2018) *Understanding the UN climate change regime.*

van der Woude, A. M., Hayami, A., and De Vries, J. (1995) *Urbanization in history: a process of dynamic interactions.* New York: Oxford University Press.

Viteri, F. E., and Gonzalez, H. (2002) "Adverse outcomes of poor micronutrient status in childhood and adolescence," *Nutrition Reviews*, 60(5 Pt 2), pp. S77–83.

Voss, R., Quaas, M. F., Schmidt, J. O., and Kapaun, U. (2015) "Ocean acidification may aggravate social-ecological trade-offs in coastal fisheries," *PLoS One*, 10(3), pp. e0120376.

Wang, H. D., Naghavi, M., Allen, C., Barber, R. M., Bhutta, Z. A., Carter, A., et al. (2016) "Global, regional, and national life expectancy, all-cause mortality, and cause-specific mortality for 249 causes of death, 1980–2015: a systematic analysis for the Global Burden of Disease Study 2015," *Lancet*, 388(10053), pp. 1459–544.

Wang, L., Southerland, J., Wang, K. S., Bailey, B. A., Alamian, A., Stevens, M. A., and Wang, Y. F. (2017) "Ethnic differences in risk factors for obesity among adults in California, the United States," *Journal of Obesity*, 2017, pp. 10.

Watkins, R., Palmer, J., Kolokotroni, M. (2007) "Increased temperature and intensification of the urban heat island: implications for the human comfort and urban design," *Built Environment*, 33, pp. 85–96.

Watts, N., Amann, M., Ayeb-Karlsson, S., Belesova, K., Bouley, T., Boykoff, M., Byass, P., Cai, W., Campbell-Lendrum, D., Chambers, J., Cox, P. M., Daly, M., Dasandi, N., Davies, M., Depledge, M., Depoux, A., Dominguez-Salas, P., Drummond, P., Ekins, P., Flahault, A., Frumkin, H., Georgeson, L., Ghanei, M., Grace, D., Graham, H., Grojsman, R., Haines, A., Hamilton, I., Hartinger, S., Johnson, A., Kelman, I., Kiesewetter, G., Kniveton, D., Liang, L., Lott, M., Lowe, R., Mace, G., Odhiambo Sewe, M., Maslin, M., Mikhaylov, S., Milner, J., Latifi, A. M., Moradi-Lakeh, M., Morrissey, K., Murray, K., Neville, T., Nilsson, M., Oreszczyn, T., Owfi, F., Pencheon, D., Pye, S., Rabbaniha, M., Robinson, E., Rocklov, J., Schutte, S., Shumake-Guillemot, J., Steinbach, R., Tabatabaei, M., Wheeler, N., Wilkinson, P., Gong, P., Montgomery, H., and Costello, A. (2018) "The Lancet countdown on health and climate change: from 25 years of inaction to a global transformation for public health," *Lancet*, 391(10120), pp. 581–630.

Wheeler, T., and von Braun, J. (2013) "Climate change impacts on global food security," *Science*, 341(6145), pp. 508–13.

Whitmee, S., Haines, A., Beyrer, C., Boltz, F., Capon, A. G., de Souza Dias, B. F., Ezeh, A., Frumkin, H., Gong, P., Head, P., Horton, R., Mace, G. M., Marten, R., Myers, S. S., Nishtar, S., Osofsky, S. A., Pattanayak, S. K., Pongsiri, M. J., Romanelli, C., Soucat, A., Vega, J., and Yach, D. (2015) "Safeguarding human health in the Anthropocene epoch: report of The Rockefeller Foundation–Lancet Commission on planetary health," *Lancet*, 386(10007), pp. 1973–2028.

Wilkinson, J. (2009) "The globalization of agribusiness and developing world food systems," *Monthly Review—an Independent Socialist Magazine*, 61(4), pp. 38–49.

Wilkinson, R., and Pickett, K. (2010) *The spirit level: why equality is better for everyone*. London: Penguin Books.

Wilkinson, R., Pickett, K., and De Vogli, R. (2010) "Equality, sustainabilty, and quality of life," *BMJ Global Health*, 341.

Williams, R. (1980) "The politics of nuclear disarmament," *New Left Review*, (124), pp. 25.

Wilson, G. R., and Edwards, M. J. (2008) "Native wildlife on rangelands to minimize methane and produce lower-emission meat: kangaroos versus livestock," *Conservation Letters*, 1(3), pp. 119–28.

Wolch, J. R., Byrne, J., and Newell, J. P. (2014) "Urban green space, public health, and environmental justice: the challenge of making cities 'just green enough,'" *Landscape and Urban Planning*, 125, pp. 234–44.

Woodward, A., Smith, K. R., Campbell-Lendrum, D., Chadee, D. D., Honda, Y., Liu, Q., Olwoch, J., Revich, B., Sauerborn, R., Chafe, Z., Confalonieri, U., and Haines, A. (2014) "Climate change and health: on the latest IPCC report," *Lancet*, 383(9924), pp. 1185–9.

World Bank (2010a) *Cities and climate change: an urgent agenda*. Washington, DC: World Bank Group.

World Bank (2010b) *World development report 2010: development and climate change*. Washington, DC: World Bank.

World Bank (2012) *Global monitoring report 2012: food prices, nutrition, and the Millennium Development Goals*. Washington, DC: World Bank Group.

World Food Program (2017) *Global report on food crises 2017*. Rome: World Food Program.

World Health Organization (WHO) (2003) *Diet, nutrition and the prevention of chronic diseases. Report of a Joint WHO/FAO expert consultation. Technical Report Series 916*. Geneva: World Health Organization.

World Health Organization (WHO) (2011) *Rio Political Declaration on Social Determinants of Health*. Geneva: World Health Organization.

World Health Organization (WHO) (2014) *Quantitative risk assessment of the effects of climate change on selected causes of death, 2030s and 2050s*. Geneva: World Health Organization.

World Health Organization (WHO) (2015a) *Global action on the social determinants of health to address health equity: supplementary report on progress in implementing the Rio Political Declaration on Social Determinants of Health and WHA65.8*. Geneva: World Health Organization.

World Health Organization (WHO) (2015b) *World malaria report 2015*. Geneva: World Health Organization.

World Health Organization (WHO) (2016a) *Ambient air pollution: a global assessment of exposure and burden of disease*. Geneva: World Health Organization.

World Health Organization (WHO) (2016b) "Global monitoring of action on the social determinants of health: a proposed framework and basket of core indicators." Available at: http://www.who.int/social_determinants/monitoring-consultation/en/.

World Health Organization (WHO) (2016c) *Global report on urban health: equitable, healthier cities for sustainable development*. Geneva: World Health Organization and United Nations Human Settlements Program.

World Health Organization (WHO) (2016d) *Urban green spaces and health: a review of the evidence*. Denmark: World Health Organization.

World Health Organization (WHO) (2016e) *World health statistics 2016: monitoring health for the SDGs, sustainable development goals*. Geneva: World Health Organization.

World Health Organization (WHO) (2018) "Backgrounder 1: The Commision on Social Determinants of Health—what, why, and how?" Available at: http://www.who.int/social_determinants/final_report/csdh_who_what_why_how_en.pdf?ua=1.

World Meteorological Organization (WMO) (2017) "WMO statement on the state of the global climate in 2017." Available at: https://library.wmo.int/opac/index.php?lvl=notice_display&id=20220#.WzpECX4nagR.

Zhang, B., Xie, G. D., Gao, J. X., and Yang, Y. (2014) "The cooling effect of urban green spaces as a contribution to energy-saving and emission-reduction: a case study in Beijing, China," *Building and Environment*, 76, pp. 37–43.

Zhao, C., Liu, B., Piao, S., Wang, X., Lobell, D. B., Huang, Y., Huang, M., Yao, Y., Bassu, S., Ciais, P., Durand, J. L., Elliott, J., Ewert, F., Janssens, I. A., Li, T., Lin, E., Liu, Q., Martre, P., Müller, C., Peng, S., Peñuelas, J., Ruane, A. C., Wallach, D., Wang, T., Wu, D., Liu, Z., Zhu, Y., Zhu, Z., and Asseng, S. (2017) "Temperature increase reduces global yields of major crops in four independent estimates," *Proceedings of the National Academy of Science of the United States of America*, 114(35), pp. 9326–31.

INDEX

Page numbers followed by *f* indicate figures; page numbers followed by *t* indicate tables; page numbers followed by *b* indicate boxes